Thinking About Technology

Thinking About Technology

*How the Technological Mind
Misreads Reality*

Gil Germain

LEXINGTON BOOKS
Lanham • Boulder • New York • London

Published by Lexington Books
An imprint of The Rowman & Littlefield Publishing Group, Inc.
4501 Forbes Boulevard, Suite 200, Lanham, Maryland 20706
www.rowman.com

Unit A, Whitacre Mews, 26-34 Stannary Street, London SE11 4AB

British Library Cataloguing in Publication Information Available

Library of Congress Cataloging-in-Publication Data

Names: Germain, Gilbert G., 1954- author.
Title: Thinking about technology : how the technological mind misreads reality / Gil Germain.
Description: Lanham : Lexington Books, [2017] | Includes bibliographical references.
Identifiers: LCCN 2017026370 (print) | LCCN 2017024802 (ebook) | ISBN 9781498549547 (Elec-
    tronic) | ISBN 9781498549530 (cloth : alk. paper)
Subjects: LCSH: Technology--Philosophy. | Comprehension. | Reality. | Thought and thinking.
Classification: LCC T14 (print) | LCC T14 .G37 2017 (ebook) | DDC 601--dc23 LC record available
    at https://lccn.loc.gov/2017026370

Printed in the United States of America

For Sheri and Emma

Table of Contents

Acknowledgments

Several years ago my fellow colleague, Ron Srigley, suggested I deliver a series of lectures to a large, mandatory first-year undergraduate class on the themes addressed in this book. Without his insistence, neither the lectures nor this book, an outgrowth of those talks, would have seen the light of day. Thank you, Ron. I also would like to acknowledge my longtime friend and confidante, Peter McGuire, whose careful reading of an earlier iteration of this work helped make the final product a better book. Emily Roderick, Madhumitha Koduvalli, Paula Williamson, and Joseph Parry, editors at Lexington Books, also deserve special mention for their care and guidance throughout the editing and production process. Lastly, I thank Scott McBride for providing the cover art, a fitting visual complement to the theme of this volume.

Introduction

Thinking about technology is not the same as thinking technologically. There is no lack of the latter these days. On the contrary, it is ever-present if for no other reason than thinking today is identified with thinking technologically, with the problem-solving mentality. It is not an exaggeration to say that to the extent we think, we think instrumentally. Understanding for us is in the service of building a better world, the nominal raison d'etre of modernity. The construction project of common cause requires that we employ our wits to build a world that conforms to our designs for it. We think to gain the requisite practical knowledge to affect the end we desire or to make the ideal real. Governments, corporations, colleges—virtually all contemporary institutions are united in the task of extending powers of control and rational management as a means of actualizing the best of all possible worlds.

This book has nothing to do with instrumental or technological thinking other than to highlight its lock on the popular zeitgeist and the consequences of this hold. But to do the latter demands we investigate the former. Thinking about technology allows us to see more clearly the nature of the civilizational project to which we are committed. To what end? Ultimately, for the same purpose Socrates spent his life questioning his fellow Athenians' views on issues of perennial importance. An unexamined life is a diminished life. There is something Lilliputian about unthinkingness. To live and work entirely within the socially constructed bounds of one's age is to limit needlessly one's horizons. True thinking, in contrast, is liberating. Socrates was a free man, despite his falling victim to the ministrations of the state, because he was not entirely a creature of the age he inhabited.

Thinking about technology aims for a similar effect. To think about technology is to gain some critical distance from a determinative force in our lives for the purpose of knowing ourselves, and the world, more fully. More to the point, to think about technology is to grapple with the force in our lives most responsible for the hijacking of thinking. So if changing the world is technology's business, ours is to reflect on the will to change and its foreclosure of critical thought. In the final analysis, to think about technology is to defend thinking as such.

If anything changes as a result of this reflective exercise, it will be less the world than our apprehension of it. Thinking about technology alters perception and it is in the realm of perception that freedom resides. Any practical consequence that may flow from this reflection remains incidental to the act itself. This is not to conclude, however, that thinking about technology is not without practical effect. One of the axioms underpinning this investigation is that the world we make reflects the world as we perceive it. Because the exterior or artifactual realm and the inner or perceptual world are reciprocally conditioned, any fundamental reorientation in perception has implications for how we go about structuring our lives. So while I am under no illusions that thinking about technology has the capacity to transform already established, broad-based cultural perceptions and behaviors, I am more sanguine about thinking's potential to loosen the technological hold in a way that might aid our maneuvering through the technological thicket. And that is a not inconsiderable start.

This small book devoted to a big idea is divided into numerous short chapters. These chapters can be categorized into three groupings. The first pertains to scene setting. Chapter 1 introduces the major themes addressed in the book through an analysis of two literary works. The second cluster, chapters 2 through 6, explores in greater detail (and via more straightforward analytical means) the previously cited themes. These chapters comprise the core of our thinking about technology. Lastly, chapters 7 and 8 constitute an exercise in what might be called "thinking past technology." Their aim is to explore post-technological ways of coping with a technological order.

A more detailed account of the above characterization looks as follows. As a framing device for the analysis that ensues, chapter 1 offers readings of Plato's *Phaedrus* and George Saunders's "Jon." These accounts highlight, in an introductory fashion, the tension between an erotic encounter with the world and a technologically mediated one.

The following chapter departs sharply from the previous. With the idea in mind that thinking about technology has to do with reflecting on our placement within a more encompassing order of things (or "world"), attention shifts to an analysis of worldviews or to variances in big-picture conceptions of the world. A distinction is drawn here between two categorically distinct modes of perception, one of which is aligned with the technological way of perceiving reality. The remainder of the chapter is given over to detailing what separates the technological worldview from a pre-technological world-picture, especially the classical Greek version. Included in this discussion of "the ethos of technology" is the transition from premodernity to modernity, technology's function as an ideology, and an investigation of efficiency as a key technological variable.

Thinking about technology takes a more critical turn in the third chapter. The pathology of the technological vision is examined through an analysis of one of modern technology's most avid and popular exponents, Kevin Kelly. It is argued that his suggestion that humanity is best served by acting in conformity with technology's wants captures perfectly the spirit of our times. It nicely encapsulates as well a misguided understanding of how humans are best served, a misunderstanding whose lineaments are explored further in the remainder of the chapter in the form of a critique of the posthumanizing thrust of technological progress.

Chapters 4 and 5 both recapitulate the preceding analysis and sharpen its critical edge through a reading of Jean Baudrillard's thoughts on technology. Baudrillard's oeuvre is sifted through to lay bare the cluster of related concepts that underpin the technological project. His idiosyncratic treatment of "evil" is a focal point in chapter 4, along with evil's significance to an analysis of technology both as idea and practice.

Simulation for Baudrillard is the "natural" outgrowth of the technological drive to remake the world as an ordered system. Chapter 5 is devoted to exploring the consequences of the technological impetus to perfect reality, which special reference to George Saunders's "Jon" as a case study. The primary effect, in keeping with the principle of evil and reversibility, is the effacing of the real by its simulated double. To live in an advanced technological order is to live within the bounds of the recreated real, in a "world" within the world. Simulation's connection to de-eroticization is underscored here as well.

In its function as a worldview and an ideology, technology constitutes a take-for-granted belief system. Chapter 6 explores further technology as a contentious faith-based paradigm of action. For comparative purposes, a counterargument is presented that holds that the efficiency principle (from which the ethos of technology draws its energy) is less an article of faith for us than a reasoned means of realizing the best practicable society. The chapter closes with a critique of the counterclaim regarding the social utility of efficiency.

Chapter 7 is the first of two chapters that move from thinking critically about technology to a consideration of what thinking past technology might mean and its importance with respect to the more general task of thinking about technology. Here this reorientation takes the form of a discussion of phenomenological insights into the body-world connection as an antidote to the "mind" fetish that pervades the technological worldview. Commentary provided by Maurice Merleau-Ponty and Alvin Noë is given special attention to this end. Continuing with the same thematic pursuit, the chapter then segues into an analysis of two seemingly disparate thinkers who happen nonetheless to share an abiding interest in human embodiment. David Bohm and Eric Voegelin are shown to be defenders of a mode of perceptual experience that presupposes a participatory or erotic encounter with reality. They are held up as exemplars of what it means to think outside the technological box, and their inclusion both secures and extends earlier claims regarding the centrality of *eros* to lived experience.

Chapter 8 is idiosyncratic. Here, thinking past technology involves thinking about the practical consequences of abandoning faith in the technological ideal. Alternatively put, this chapter tries to answer this question: How might it be possible to keep *eros* alive in a world that conspires against it? The response takes the form of rough guidelines for action. These rules of thumb are offered as practical suggestions for navigating the contemporary landscape in ways that bypass technology's helping hand.

The book's concluding chapter rounds out an analysis of technology by considering the problematic nature of the "more is better" ethic aligned with the technological ideal. Re-impressed is the claim that thinking about technology is ultimately a defense of thinking itself.

On a final note, it merits mentioning that this study, in its present form, evolved out of a series of lectures I delivered to a large class of undergraduate students of varied academic interests. It is my hope that the tone of this inquiry still evokes the spirit of its origin.

Chapter One

Open and Shut

As a point of departure, let us consider a time and place long before the technological era. Let's enter a world that did not situate at its center the technological imperative and the accompanying notion that "man is the measure of all things."[1] While there are many such eras, an especially attractive one for our purposes is the world of ancient Athens. And within this general setting, let's draw our attention to a discourse that focuses on the question of the relationship between humans and those "things" to which human are related that collectively may be called "nature." I am thinking here of Plato's *Phaedrus*.

The "trees and open country" affect the tone and the content of the *Phaedrus* in a way that belies Socrates's claim in the dialogue that nature is incapable of edifying. The self-professed "lover of learning" claims only "men in the town" are capable of teaching him anything.[2] It should be noted, parenthetically, that Plato's assessment of Socrates's rejection of natural philosophy and his embrace instead of the pursuit of "human wisdom" is likely overstated.[3] For one, we know, on Aristophanes's account in *The Clouds*, that the comic playwright lampooned the middle-aged Socrates's penchant for analyzing the natural order, a jab that would not have played to the crowd if its portrayal of Socrates missed the mark entirely.[4] It is true that, in Plato's *Apology,* Plato has Socrates reference Aristophanes's claim regarding Socrates's affiliation with natural philosophy only to disavow it.[5] Yet the rebuke in itself does not disprove the association between Socrates and natural philosophy. Setting aside reasons why Plato may have had an interest in sharply distinguishing the human and natural domains, it is conceivable the political upheavals Socrates witnessed in his lifetime forced upon him an intellectual reorientation. Arguably, the calamity that was the Peloponnesian War precipitated a spiritual crisis that in turn led to the soul-searching ways of Socrates and others. In this reading, the human wisdom Socrates came to respect and pursue stands as a complement to the study of nature, not a radically separate branch of understanding. The following analysis aims to illustrate this point.

The only Platonic dialogue explicitly set outside the city of Athens, the *Phaedrus*, opens with a chance meeting between Socrates and Phaedrus at the edge of town. Socrates learns that his friend has just had a long conversation with several companions that morning and is now, on doctor's orders, on a walk through the countryside. Phaedrus recounts his physician saying that open air perambulations are more "invigorating" than treading the colonnades of the city.[6] The identification of vigor—of energy, or the life force—and the outdoors is marked. It also warrants mention that the dialogue begins in the late morning and proceeds through the afternoon, a time of day when the sun—the symbol of life and intelligence—is highest and therefore most vigorous.

On their walk Socrates queries Phaedrus about the topic of conversation that earlier occupied his friend and learns it was the subject of love. Phaedrus recalls that Lysias had given a long and eloquent speech on love and finds Socrates so taken with the idea of such a speech that he is called upon by Socrates to recite it. The extent of Socrates's enthusiasm is revealed in his telling Phaedrus he is willing to walk to Megara and back if need be to hear the entire speech, a journey of over fifty miles. Phaedrus scoffs at the suggestion that he remembers Lysias's speech verbatim, but agrees to provide Socrates with a sketch of its main points.[7] Observant Socrates has other plans. Thinking all along that Phaedrus is more acquainted with Lysias's discourse on love than he lets on, Socrates gets Phaedrus to reveal a transcript of the speech that Lysias had concealed beneath his cloak.

With the script in hand, the prospect of a leisurely walk through the countryside comes to an abrupt end. The act of reading requires a level of concentration that only stillness can accommodate. The spoken word has an affinity with improvisation and spontaneity, with fluidity and movement, in a way the fixed written word does not. So the interlocutors look for a place to rest. Socrates suggests they wander off the path and toward the banks of a nearby river. At Phaedrus's suggestion, the barefooted friends make their way to a shady riverside spot under a tall plane tree or *agnos*. The symbolism attached to this resting place is revealing. Approaching the noon hour and in the middle of the summer, when the sun and heat are at their most intense, two friends settle down to talk of love under a species of poplar tree (in full bloom, yet) that in the ancient world was known as the "chaste tree." According to local lore, the seeds of the *agnos* were thought to possess anaphrodisiac properties and were used to keep the allures of the flesh at bay from the well-intentioned, allowing them instead to focus on divine matters. Interestingly, and perhaps somewhat paradoxically given its association with chastity, the seeds of the plane tree also were thought to facilitate birthing.

Regardless, having settled into their pastoral enclave, Socrates immediately responds to the sensory enchantments of the surroundings. He comments eloquently on the fragrance of the flowering *agnos*, the coolness of the stream, the welcoming shade of the overhanging tree, the freshness of the air, the "shrill summery music of the cicada choir," and the softness of thick grass below. He then thanks Phaedrus for being "the stranger's perfect guide" to the countryside, to which Phaedrus curtly replies, "Whereas you, my excellent friend, strike me as the oddest of men."[8] Socrates is thought a strange man because he seems out of place in the elements. By his own admission, the land beyond the city walls is alien territory to Socrates, a conclusion Phaedrus appears to have reached independently given the reaction of his friend to the environs.

At this point Socrates "apologizes" for his unfamiliarity with the countryside, in both senses of the term. A gentleman, he asks to be forgiven for his strange behavior, but then he proceeds to defend his response by arguing that as a lover of learning he is drawn toward the city and the men who live there. But can Socrates be taken seriously here? There is no doubting he was a man of the city, but is Socrates not being ironic in claiming that he is a stranger to the natural order? Clearly, Phaedrus's suggestion that Socrates had virtually no acquaintance with the world outside the walls of Athens is overplayed. We know, at the very least, that the historical Socrates fought in the Peloponnesian War and that his military adventures took him outside his hometown for extended periods of time. Also, it stretches credulity to believe that one could live in an ancient Greek "city" and be completely unversed in matters pertaining to "trees and the open country." Even Socrates's physiognomy and general bearing suggest he had a good deal of the rustic in him and that despite his intellectual sophistication, he was anything but urbane or effete.

Rather than taking the claim regarding Socrates's naiveté in rustic matters literally, we must consider that Plato consciously employed references to the natural setting of the *Phaedrus* to suggest symbolically the ontological unity of the human and natural orders. To further explicate this union requires the birth of an idea, one that pertains to the notion of love, albeit a chaste love. So it is that protected from the sun directly above by the cover of a tree associated with natality, Socrates lays down and prepares to give birth. His telling Phaedrus that it matters not to him what position he takes during his recital is not a mere casual remark since it is not Phaedrus (Lysias's mouthpiece) who ultimately is to deliver the word.

But of course Phaedrus is the first to deliver. After politely praising Lysias's text, as read by Phaedrus, Socrates is coaxed by Phaedrus to offer his true opinion of it. Socrates, who earlier admired the rhetorical qualities of the speech, admits he has heard better disquisitions on love than Lysias's, although he remains reluctant to address the issue himself. As in the opening of the *Republic*, Socrates's silence ends only when the threat of force is invoked. Only this time what compels Socrates to speak is not the threat of physical force but Phaedrus's threat to never again speak about speeches. Socrates praises Phaedrus's cleverness in finding a means to compel "a lover of discourse" to speak, yet the praise seems insincere being directed toward a mere reciter of another's speech of dubious philosophical merit.[9]

So, with a fairy tale flourish, Socrates begins to speak to the question originally addressed by Lysias, namely "whether one should preferably consort with a lover or a nonlover."[10] In typical fashion, Socrates argues that in order to address this question it is first necessary to understand the terms of the debate, the primary one being the meaning of love. With this move he casts aside the more conventional concerns voiced in the original exposition and enters the realm of philosophical investigation. Not for the first time, Socrates in the *Phaedrus* equates love with desire, and says that love in the everyday sense of the term is a strong passion directed toward bodily beauty. Love, the innate desire for bodily pleasure, is called by Socrates a "ruling or guiding principle that we follow."[11] But it exists alongside another such principle, one he calls "an acquired judgment," that rationally aims at what is best.[12] So there are in effect two types of love, and Socrates plays to the standard philosophical bias by pitting the debased and irrational love of sensual pleasure against the higher and rational desire for right conduct.

This portion of his analysis complete, Socrates half-mockingly asks Phaedrus if he believes him to be "divinely inspired," on account of the eloquence of his speech.[13] Immodestly, Socrates says he thinks he is, and Phaedrus agrees. Socrates then uses this moment to silence his interlocutor on the grounds that there appears to be "a divine presence in this spot"[14] to which the philosopher is evidently attuned. Clearly, Socrates at this point in the dialogue plays up the conventional association between things divine and eloquent, or inspired, speech. Inspired speech is superhuman speech, speech that appears to emanate from a source beyond the merely human. The assumption here is that mere mortals are largely incapable of grandiloquence. But Socrates knows better. He realizes that such speech is within the power of mere humans to utter. And, more importantly, he knows that the non-philosophic crowd tends to conflate grandiloquence with insight, to misleadingly identify beautiful speech with true speech.

Socrates is fully cognizant of Phaedrus's conventionalism. But rather than trying to upend it, he instead plays to Phaedrus's "common sense" understanding. In other words, he knowingly adopts the mantle of the "divinely inspired" in order to convey to Phaedrus, in a manner appropriate to his abilities, a *philosophical* understanding of love. And against type, this philosophical portrayal of love praises its irrationality or madness. Contra Lysias, love, per se, is not an evil according to Socrates. There is no denying that the madness of love can result in much confusion and misfortune, especially when the object of love is another human being. But, Socrates adds, "the greatest blessings" are said to be bestowed upon those who are taken by a madness that is "heaven sent."[15]

Socrates defends the claim that those possessed by a divine madness have conferred upon them the greatest blessings by examining the nature of the soul, both human and divine. The universe or cosmos for Plato is a living entity, and its life is evidenced in its motion. An immobile, static universe would not be a universe, or at least not the universe of human experience, for both the human and cosmic realms exhibit movement and therefore appear informed by some kind of life-principle. For Plato this life force or *anima mundi* is "soul," and soul is defined simply as the power of self-motion. For Plato, then, anything capable of self-movement has soul in it. Moreover, as self-moving, soul has no beginning, since such a beginning would imply that the power of self-motion was imputed to soul by a force external to it, in which case soul would lose its status as self-moving. The soul equally has no end, Socrates says, since "it cannot abandon its own nature."[16] Thus all things capable of self-movement are possessed of deathless soul. The sun that crosses the heavens, the stars that traverse the night sky, the goats that graze the Attic countryside, and the goatherd that tends to them have something of the immortal in them. They may inhabit differing orders of being, but they are one in the possession of soul.

While depicting this holistic vision of the cosmos, Socrates must also account for differentiation within the unity of being. Not all things participate equally in the unity of being, for clearly there are meaningful differences between a blade of grass, a human being, the celestial sky, and the gods. In the *Phaedrus*, Socrates provides a mythological account of this ontological hierarchy through the imagery of the so-called Chariot Allegory.[17] The allegory portrays all self-moving beings as ensouled entities, beings composed of both soul and body. However, the spark of divinity that all such beings share is not distributed equally. The admixture of soul and body varies according to the type or nature of a being. The higher the ratio of soul to body possessed by such a type, the closer it is to divinity. Human beings are unique earthly creatures in that differentiations in the soul-body ratio exist *within* the species. For Plato, persons possessed by heaven sent madness are those for whom the divine soul most fully illuminates the human soul. Such a person, of course, is for Plato the philosopher, someone who more than any other human type is blessed by the presence of the divine within him.

It is in the *Timaeus* that Plato explicates most fully the ontology of soul and its relation to the created cosmos.[18] As mentioned, the cosmos or visible universe is for Plato ensouled and therefore the realm of becoming or movement. All the constituent elements of the cosmos are admixtures, of varying proportions, of matter and soul. In the *Timaeus* Plato depicts the creation of the cosmos in terms of the imposition of soul upon matter, or *psyche* upon *soma*. For Plato, the soul that does the imposing is not a blind force but an immaterial substance illuminated by *Nous*, Mind, or Reason. What is illuminated by soul is the paradigmatic form that is the eternal world, or the Idea of the Forms in the language of the *Republic*.

The creative force that impresses form upon matter Plato calls the Demiurge (*demiurgos*). The Demiurge is said to fix the image (*eikos*) of the divine paradigm upon a preexistent material substratum. The act of imposition is necessary because the material substratum of the cosmos constitutes a distinct order of being with its own movements. The material order tends to resist the imposition of form upon it, but clearly the existence of the cosmos is evidence that order, or at least some measure of order, has prevailed over complete disorder or utter randomness.

Plato's depiction of the cosmos suggests the universe to be a symbol of its eternal ground. The cosmos is a moving, living, concrete image of a divine and eternal intelligence. As are all images, the cosmos simultaneously is and is not that which it represents. The cosmos implies the eternal world. It is an explicit manifestation of an implicit order, an unfolding of the eternal into the realm of space and time. In this sense the cosmos is a world of becoming that bears the imprint of eternity, a realm of being-in-becoming.

The cosmos is intelligible for Plato to the extent that like knows like. Humans can comprehend the order of the whole because, as an image of the divine soul, the human soul is attuned to the source of cosmic order. But for all this we cannot understand the divine mind fully (as can the Demiurge) since humans are composite beings whose material substrate necessarily resists to some degree the imposition of the eternal or paradigmatic form.

What Plato's holistic teaching reveals is the impossibility of "objective" knowledge of the cosmos. Reality for him is not a dead "thing" that we gaze upon as if from without, but a living organism of which we humans constitute a part, albeit a privileged part. Whatever knowledge of the whole we gain is gathered in the context of our participatory relationship with it. We come to know the whole from the "inside," as it were. So, on the one hand, Plato's cosmos is a welcoming place to the extent humans share, with all created reality, soul. Yet, on the other, our position and status within the whole means for Plato that our grasping of reality's ground will forever remain allusive. While we may gain sight of the divine intelligence, we do so only through the mask of its representations.

Two features of my treatment of Plato's analysis are especially germane to our discussion and therefore merit underscoring. The first is the depiction of the universe as enchanted. A pre-modern, Plato rejects the idea that the cosmos is mere matter in motion and that mind is a mere accident of the operation of the laws of physics and chemistry. This disenchanted view is ours and is held widely by the modern scientific community for good reason.[19] Admitting that the cosmos might be constituted in a way to produce consciousness would say something about the nature of nature that subverts a central underpinning of the modern scientific worldview. It should be made known, in this regard, that Plato continues to have allies into the twenty-first century. Thomas Nagel, for one, argues forcefully for an alternative to materialist naturalism, for a view that holds there is a "cosmic predisposition to the formation of life, consciousness, and the value that is inseparable from them."[20]

If the universe is panpsychic, and therefore predisposed to giving birth to mind, then arguably the mind to which it gives birth is likewise disposed to experiencing the ground from which mind emerges. It is precisely because mind is an expression of the cosmic order that mind remains open to that order, at least potentially. What is more, this openness is erotically charged. That is to say, mind is existentially perturbed by virtue of its capacity to sense its ground as existing beyond itself. Because mind, according to this reading, is a cosmic irruption, the experience we call consciousness reverberates within a context that transcends its own delimitation as a conscious entity. Consciousness, to repeat, is seen here as a participatory experience. Mind, for all we know, might be an exceedingly rare phenomenon, but that doesn't render it unnatural or disconnected from the world that gave it birth. It is for this reason that human consciousness is, arguably (and as Plato suggested), an integral part of the cosmic order and as such responsive to this overarching order.

We are brought now to the topic of erotics. Most famously, Plato has Socrates say in *The Symposium* that he understands nothing other than "*ta erotika*,"[21] or things pertaining to *eros*. This is a strange claim on many levels, none more than that it emanates from a self-professed philosophical skeptic. In any event, we must take this assertion of Socrates's seriously. If Socrates knew anything it was that he was a lover, and he understood that his hungering for wisdom was not self-generative. The search for knowledge is not, at bottom, a deliberative act. Rather, it is a response to a calling, an impulse from within generated by an existential condition not of our making. This is Plato's insight into *eros*. We are, in his words, a "plaything of the gods,"[22] players in a cosmic drama the meaning of which we are not entirely privy.

Plato incorporates the human drama within a broader cosmic setting as a means of articulating the core of the human experience. Human consciousness, for him, participates in an order that transcends the merely human. With due attention, he seems to suggest, we realize the extent to which being human is being open to the world beyond the simply human. To put a label on it, we could call this orientation a "participatory worldview."

The purpose of my outlining the contours of a largely defunct reading of reality is to highlight, by way of contrast, our own interpretive stance. This is the first step in bringing into relief what lends our world its idiosyncratic character. Step two requires another imaginative repositioning. We venture now from Plato's midsummer ruminations about love and the cosmos to an allegorical love story as conveyed by George Saunders in his short story, "Jon."[23]

As noted, the mise-en-scène of the *Phaedrus* is the great outdoors. This is true both literally and figuratively. The discourse about the nature of love and desire that unfolds out in the open itself is pervaded with the symbolism of openness. This openness is revealed in the etymology of the verb *desire*, with its celestial link to *de sidere* (of, or from, the stars). In desiring, we wish for or await what the stars will bring: We are open to what is not but what we wish to be. Desiring beings are therefore displaced beings. Socrates may have conceived himself as a creature of nature, but he understood that human nature is such that it holds within it a dimension that grounds it in something that transcends simple creaturely existence. What this "something" is is not of concern at this moment. Suffice it to say that from a Platonic perspective, to be human is to reside in between mere animality (or life within the bounds of animal needs) and godliness (or life of fulfilled desire). In short, to be human is to yearn for something one does not have and cannot ever have—the repose that comes with total self-satisfaction. This unease is a given and is both a blessing and a curse since it defines what it means to be human but provides a definition that informs us self-satisfaction is not our lot.

The same cannot be said for us moderns. A curse rather than blessing, the erotic pull that draws us out of a state of complacency is deemed a problem in need of remediation. That, at least, is the hypothesis upon which the following study is premised and a guide in our ruminations on technology. *Eros* for us is an evil that must be extirpated. Being "in want" is anathema to the proper functioning of the social order. Anything that smacks of existential openness is contrary to the spirit of our times. It is for this reason that the air of openness that suffuses the *Phaedrus* is utterly absent in George Saunders's "Jon," at least initially. Through the lens of the short story's eponymous protagonist, Saunders supplies us with a picture of a closed universe whose end is the production of equally closed, unerotic, or self-satisfied beings.

The airless quality that hangs over the narrative is exemplified in its setting, a compound called the "Facility" in which the plot almost entirely unfolds.[24] The Facility (one of many identical compounds, we are led to believe) occupies a placeless space, somewhere in the American Midwest. No site-specific features mark its location within a broader worldly context—no plane trees by the river here—further accentuating its hermetic character. The same goes for the story's location in time. Never specified, "Jon" unfolds in what appears to be the near future. From the outset, then, one is struck by the featurelessness of the story's setting. The backdrop is as generic as those big box stores strewn across the contemporary suburban landscape.

The "world" of Saunders's "Jon" is blandly uniform by design. The architects of Jon's world work assiduously to ensure nothing ever goes wrong. Since things "go wrong" when desires go unfulfilled, Jon's world is a place where energies are singularly devoted to satisfying desires. To the extent this ambition is met, Jon's world is desireless. His life is arranged so as to produce self-contentment, as is everyone else's life in the Facility. And what Leo Tolstoy once said of families is equally applicable to individuals: All happy persons are alike, and all unhappy persons are unhappy in their own way. A de-eroticized environment is by definition a world of homogenized experience.

In "Jon," Saunders addresses this phenomenon with a sly take on modern consumer culture. It warrants mentioning at the outset that "Jon" works as a story because the reader is encouraged to realize that the variances between our modern consumer culture and the one depicted in his story are differences without a distinction. The same principles inform both. The same ends are operative in both. The only difference is that Jon and his friends live in a social order that realizes this shared ideal more effectively than we do today.

Key to my argument is the contention that this ideal is the fantasy that animates the technological project. It is the dream of perfect or total domination, whereby through human intellection and ingenuity the world loses its status as an object and is rendered wholly compliant to the controlling will. The image of perfection that technology chases is a world that functions in the manner we want and does so efficiently. What makes us distinctly modern is precisely the expectation that the world, both human and nonhuman, can be subjected to effective, rational control, or operationalized. The modern spirit is coterminous with the spirit of technology. And this spirit is totalizing. Nothing is seen as immune to being operationalized. From supply chains to chainsaws, everything is treated as a system whose functionality is to be maximized.

In business circles, the spirit of technology reveals itself most unequivocally in the phenomenon known as "total quality management" or TQM.[25] TQM is an extension of Taylorist principles of scientific management as applied to economic production. These principles pertain not only to the efficient production of quality goods and services but to their distribution and consumption as well. Under TQM, the entire production-consumption chain is regarded as a system whose efficient functioning is to be maximized. Today, this maximizing is effected in large part through the acquiring and processing of data generated within a given system. One preferred means of obtaining such data is the solicitation of consumer feedback. By filling out online satisfaction reports and their surrogates, consumers provide businesses enterprises with the kinds of information that help them become more efficient, less wasteful enterprises. These enterprises wish to produce what the consumer wants to consume, no more and no less. While producing undesired goods and services is clearly wasteful, so is the over- and under-production of consumables deemed desirable. The information collected from consumer feedback is employed to mitigate waste, or to "rationalize" the production-consumption dyad. The better able businesses are at rationalizing their affairs, the happier we are as consumers and the more profitable these enterprises become.

In "Jon," the consumer model that drives our society is reimagined at the next level of efficiency. Rather than rely on the vagaries of consumer compliance to do the job, we see in "Jon" the professionalization of consumer product assessment. At bottom, the Facility is a product assessment machine, an organized system that does what we already are doing, only more effectively. It procures information and in so doing provides a function endemic to any and all technologically driven cultures: it seeks to minimize inefficiencies within an established management system. So, in broad terms, the society Saunders depicts in "Jon" is just like ours. It aims to make satisfied consumers out of human beings. It wants to produce a world that works, a tensionless world with few or no unwanted surprises or frustrations. Read this way, "Jon" represents less a dark alternative to our own way of life than a purer expression of it.

As if to underscore its contiguity with the contemporary world, "Jon" unfolds as a formulaic love story. The "boy meets girl, boy loses girl only to reunite" trope is replayed here. It is the lo-fi sci-fi context within which the love affair between Jon and Carolyn emerges that lends the story its pathos. But the setting of the romance is not simply an add-on that gives the romance its color. While the love story is central to the overall narrative, it cannot be properly appreciated outside of the context in which it unfolds.

The reason why it cannot is because the world inhabited by Jon and his coworkers is programmed to extinguish *eros*. Jon and Carolyn's love story takes root within an environment inimical to love. The poignancy of the story emerges from the tension between what their love for each other demands of them and what their socialization encourages them to cultivate in terms of loyalties and attachments. The power of erotic yearning stands in stark and uncompromising contrast to a world where everything is treated as an object of actual or potential control.

Saunders wastes no time highlighting the deeroticizing impact of a technological milieu. In the short story's opening sentence, Jon recounts an instructional video where he and his friends are encouraged to watch, entitled "It's Yours to Do With What You Like!" In it, he says, "teens like ourselfs speak on the healthy benefits of getting off by oneself and doing what one feels like in terms of self-touching."[26] Jon makes a point of stressing that while the video acknowledges love may be a "mystery," it teaches that the "mechanics of love" need not be.[27] Relegating the mystery of love to the dustbin of inquiry, love qua masturbation is treated as an operation whose optimal functioning requires technical training.

The initial identification of the putative "mechanics of love" with masturbation is not serendipitous. Metaphorically speaking, the entire social order to which Jon contributes is an onanist's playground. After all, its central organizing principle is the satisfaction of personal desire, with the Facility being a means to such satisfaction. Everything in Jon's world conspires to create perfectly functioning delivery systems whose ultimate objective is personal gratification, sexual or otherwise. *Eros* is its enemy.

The paradisiacal quality of deeroticized life in the Facility is equally evident. Jon and his fellow "assessors" live in an updated version of the Garden of Eden. They inhabit a consumer paradise where they get to keep the cool stuff they rate and become icons of popular taste in the process. The assessors are product-testing rock stars and, like intramundane gods, they are buffered from slings and arrows of mere mortal existence. There is Aurabon, the drug of choice, to help erase any present and future psychological distress their conditioned existences fail to keep at bay. And the hauntings of events past are replaced with a Memory Loop that resolves by artificial means any tensions that may compromise their present-day happiness.

The "idyllic" life of the assessors comes at the cost of its utter artificiality, however. Their physical environment, for one, is wholly constructed: nothing natural intrudes, save the bits of unreconstructed reality visible through the Facility's lone observation window. Likewise, their social existence is lived out entirely within the froth of a marketing culture, where ad campaigns constitute the sum total of the assessors' imaginative landscape.

Now, a story about paradise is boring because perfection is boring. Boring places are places with nothing to do. They are places devoid of the possibility of action because true action springs from dissatisfaction with the given, and paradise of necessity is a place where one is at home with the world as it is. It is precisely desire for what is not, or for what one does not have, that moves us to act. The self-satisfied creatures of the Facility are desireless in their contentment. They function but are incapable of acting.

To get a story worth telling going, something must happen to break the seal of the perfectly functional world of the assessors. The existing condition of low entropy needs disrupting. Disorder must enter into the picture, and its entry requires energy. In "Jon," this energy takes sexual form, only this time the energy results in the carnal union of two of Jon's friends, Josh and Ruth. The sexual escapade is launched with Josh squeezing through a space separating the girls' from the boys' living quarters. In describing Josh's movement as "snakelike," Saunders associates Josh's advance with the symbolism of the serpent and its Old Testament linkages to evil and chaos.[28] Josh, it may be said, is a vector that infects the compound with a force antithetical to its ordering principle.

Not unexpectedly, what results from the act of sexual union is a real baby, Baby Amber. A real act springing from true passion results in the chaos of a newborn child. Baby Amber is evil personified because, as with every infant, she is the embodiment of unpredictability. Nothing is more antithetical to the spirit of perfect functionality than a capricious and prodigious shit machine.

As action begets action, Jon follows Josh's lead by slipping one night into his paramour's tent. Jon describes his tryst with Carolyn in the following manner:

> And though I had many times seen LI 34321 for Honey Grahams, where the stream of milk and the stream of honey enjoin to make that river of sweet-tasting goodness, I did not know that, upon making love, one person may become like the milk and the other like the honey, and soon they cannot even remember who started out the milk and who the honey, they just become one fluid, like this honey/milk combo.[29]

This passage is at once both touching and disturbing, at least from the reader's perspective. For Jon, the limited range of his powers of self-expression is not problematic. His articulation of the experience of lovemaking in the idiom of a cereal advertisement is the only kind of imagery he has got to work with and, given the material, Jon's powers of expression are more than adequate.

If there is something disturbing about this passage it is how Saunders uses language to underscore the general point that there is no strict separation between our inner thoughts and feelings, on the one hand, and the world with which we interact, on the other. In short, our inner or private world is shown never to be entirely our own. Our ability to articulate thoughts and feelings—and to some extent even to have them—is shaped both by the nature of the social order we inhabit and by the character of our relations with this order. It would be odd indeed, for instance, if living in a world given over to branding did not tend to elicit the production of truncated thoughts and caricatured feelings. And perhaps more important, and unsettling, nor should it be assumed that restrictions of this sort are recognized as such by language users. To use Platonic imagery, a significatory cave is not perceived as limiting if the cave constitutes one's cosmos.

The reference to milk and honey in Jon's post-coital soliloquy is an instance where Saunders appears to play with signification to deepen our appreciation of the underlying tension between *eros* and technology. It is in Book of Exodus that Israel is referred to as "a land flowing with milk and honey," a phrase intended to signify its overflowing abundance.[30] Here it appears Saunders is aligning love/sexual union with abundance or with what Paul Feyerabend has called "the richness of being."[31] If so, then a world without love amounts to an impoverished reality, a faint imprint of its native fullness.

As with Josh and Ruth, Jon and Carolyn's sexual union results in pregnancy, which in turn leads their marriage and what Jon describes as "the best day of our lifes thus far for sure."[32] But as in real life, the good comes with the bad, and tragedy soon strikes with the death of Baby Amber. Everyone connected with her is affected deeply by Amber's death, and Jon is no exception. "This sucks, this is totally fucked up!" he blurts in response.[33]

This decidedly less-than-eloquent lament is nonetheless an expression of real emotional pain, and consequently it cannot stand. Grief is an unproductive emotion. In a culture hell-bent on solving problems, despair is seen as an inefficiency demanding corrective attention. So Jon's grief is pharmaceutically exorcized as a means of restoring his capacity to function in the manner he was assigned. As the label suggests, Aurabon works to restore the good mood required of productive activity, and the drug performs well as a function-enhancer. No sooner is Jon on his meds than he remarks on his improved capacity to perform assessments, resulting in his winning a regional prize in assessing excellence.[34]

Amber's death has a very different long-term effect on Carolyn. Because her pregnancy prevents her from taking Aurabon, the emotional impact of Baby Amber's death lingers. Carolyn stays real. And because she does, the discrepancy between her all-too-human response to a loved one's death and that of her chemically altered cohorts becomes untenable. "Wake up and smell the coffee," she admonishes, "you feel bad because a baby dies, how about honoring that by continuing to feel bad, which is only natural, because a goddam baby died, you guys!"[35] And then, one night soon afterward, when Carolyn's *in utero* child kicks, the full import of her rejecting her posthuman environs makes itself known. "Don't worry, angel," she announces, "Mommy is going to get you Out."[36]

By "Out" Carolyn means leaving the Facility. The outstanding issue at this point is whether Jon will accompany her. Because Jon loves Carolyn, he knows his fate is tied to hers in one way or another. Jon is not all about Jon anymore. He senses as much, but he is not enamored with the prospect of going "Out" if for no other reason than he likes being "In." Paradise might be boring, but for Jon it is an especially fun kind of boring. In contrast, the thought of taking on a loser job, like working in a lumberyard, is considered by him a fate worse than death.[37]

But there is another problem with life on the outside that is cause for consternation. Jon describes the dilemma with a charm all his own:

> Of what will we speak of? I do not want to only speak of my love in grunts. If I wish to compare my love to a love I have previous knowledge of, I do not want to stand there in the wind casting about for my metaphor! . . . [I]f I want to say Carolyn, LI 34451, check it out, that is how I feel about you—well, then, I want to say it![38]

Jon is sincerely worried about the impact leaving the Facility will have on his powers of self-expression. The well-being of his inner, emotive life is as much a concern to him as that of his external, material circumstances. Jon prides himself in his ability to convey his love for Carolyn in the imagery of a RE/MAX ad. His powers of articulation are dependent on his retaining the cultural references from which he draws insight. Losing these references means losing his voice, which explains why Jon believes that should he leave the Facility, he will be reduced to communicating in grunts. Of course, the irony is that Jon already is speaking in grunts. He fails to realize that only by leaving the Facility and "casting about for my metaphor" does he stand a chance of being better able to express his love for Carolyn.

Jon's ongoing ambivalence toward life on the outside is contrasted sharply with Carolyn's resolve to leave the Facility, upon which she promptly acts. Her absence from Jon's life in the compound only serves to re-impress upon him the depth of his love for her. This longing to reunite supplies him with the determination to make the move. Yet, Jon remains to the end uncertain about what he sees of the outside world, as viewed from the Facility's observation window. The people he surveys milling about strike him as anything but an uplifting bunch. The drab and "bummed-out-looking guys in the plainest non-designer clothes ever" are an ornery lot whose default mode of communication appears to be caterwauling.[39] Much like the proles in George Orwell's *Nineteen Eighty-Four*, Saunders portrays real people as flawed and incomplete beings.[40]

Reservations aside, Jon's first exposure to the outside world is occasioned when a Facility coordinator opens a door, allowing him to peer out. He describes the experience in the following manner: "Looking out, I saw no walls and no rug and no ceiling, only lawn and flowers, and above that a wide black sky with stars, which all of that made me a bit dizzy, there being no glass between me and it."[41] It is important to remember that, at this moment, Jon is not yet out of the Facility. He is merely peering out from within the confines of his domicile. In a way, the real world at this point remains for Jon a mere picture of reality, as framed by the opening in the Facility wall. Still, the impact of this windowless image of reality is jarring and disorienting, a testament to the richness of being that has been forsaken within the Facility.

Jon then is gently pushed out, not unexpectedly given his trepidation over the outside world. He continues: "And I don't know, it is one thing to look out a window, but when you are Out, actually Out, that is something very powerful, and how embarrassing was that, because I could not help it, I went down flat on my gut, checking out those flowers."[42]

It ought to be underscored that for Jon the distinguishing characteristic of the outside world is not limited to a particular feature of it, whether it be the night sky or flowers or the grass beneath his feet. It is the being of reality itself that captures his attention. The world exerts an irresistible force upon Jon: he cannot help but be drawn toward it. The seductive power of reality is so overwhelming, it turns out, that Jon expresses shame for having been laid prostrate, a fitting emblem of his reorientation toward the being of the world. Not inert information waiting to be scanned by a probing and dispassionate eye, Jon's fascination with the flowers is in response to the call of the world itself. The world, in short, is experienced by him as erotically charged or an order of love. Jon's attraction to Carolyn can be seen as an analog of his attraction to the world at large. He experiences for the first time what it means to participate in this order, to play an active role in its unfolding, in contrast to the distancing, technological orientation, which takes the world to be a resource best fitted for controlling.

The "kickass outside yard," however, is not the culmination of Jon's existential enlightenment. It reaches its pinnacle with his sighting of Carolyn, who he initially fails to recognize on account of her plainness. But that soon changes: "to tell the truth, even with the DermaFilled neckhole and nada makeup and huge baby belly, still she looked so pretty, it was like someone had put a light inside her and switched it on." Paralleling another scene from *Nineteen Eighty-Four*, where Winston waxes poetic about an objectively less-than-beautiful woman, Jon is entranced by the inner beauty of the woman he loves.[43]

Jon's love for Carolyn is not sufficient to quell entirely his reservations about life on the outside. Intellectually, his doubts persist to the very end. What has changed, however, is Jon's sense that life beyond the confines of the Facility possesses a gravity not found within its walls. This weight is a consequence of his participating in an order of being marked by an open-endedness, in direct opposition to the closed and hypermanaged confines of the Facility. The world is *not* Jon's to do with what he likes: it acts on him as much as he acts on it. Jon's reorientation, his *periagoge*,[44] is revealed in this passage through his conflating "I" with "we." He now knows his fate as a person is not wholly self-determined but tied to that of another person. Similarly, Jon knows their future lives together will be open to the contingencies of a world not entirely of human making. It is precisely not knowing what the future holds for them that supplies him with the courage to give life on the outside a try. It is the fortuitousness of real existence that lends life its meaning, as precarious as this meaning may be.

As stated at the outset, "Jon" presents us with a distilled image of modernity. This image brings into relief and satirizes our consumer-obsessed lives. As a result, "Jon" reads us as much as we read it. Or at least it should. For in assessing the assessors' lives, we are assessing our own, which resembles Jon's life more than many of us may have the courage to admit. We inhabit our own version of the Facility. We know more about the Twitterverse than the universe, more about the antics of a trending pop star or how much a top-grossing film pulled in on the weekend than about the world under our noses. And in an age of text messaging, to what extent are our "communication skills" all that distinguishable from those of Jon's? Do we not grunt in our own uniquely re-animalized way? The parallels continue regarding the role pharmaceuticals play in our lives today. The medicalization of basic human emotions and conditions (i.e., grief or shyness) is a well-entrenched and documented practice that gives over our psychic well-being to systems of control no different in kind than those used to effect any other kind of efficiency. We have our Aurabons, and they go by a long and growing list of names such as Xanax, Cymbalta, and Prozac. And in the love and sex department, is not finding a mate or a hook-up becoming indistinguishable from shopping for one? The growing popularity of online dating sites attests to the powerful sweep of the tick-the-boxes, consumerist mentality throughout the contemporary social order.

So the bad news is that, in these and innumerable other ways, our lives are becoming as creepily surreal as the lives of the assessors depicted in "Jon." We are in the midst of exchanging the richness of being for simulations of reality whose only redeeming feature is their functionality. The collective desire to build a world that works is more powerful an idea and ideal than any alternative to it. Yet love, or *eros*, remains the only countervailing force to the push to operationalize.

The good news, Saunders proffers, is that erotic love can never be entirely suppressed, no matter how hostile an environment might be to the spirit of love. In this regard, he appears sympathetic with a Roman poet's observation regarding the futility of efforts to eradicate nature.[45] What was the erotic attraction between Jon and Carolyn if not a manifestation of the revenge of the repressed, one that informed them that their prepackaged existences in the Facility were hollow and phony? Their love for each other—their natural yearning for completion—led them to understand that there is more to life than the self, its security, and its satisfactions. They came to understand that real life is, by its nature, open-ended, and for that reason fraught with uncertainties as well.

All totalitarian regimes, real or imagined, past or present, have as their primary objective the erasure of desire in all its forms for the very reason that desirous beings are existentially unsettled. The leaders of these regimes realize that the key to total domination is the creation of self-satisfied beings. Yearning for something that transcends the given is anathema to the stability of the totalitarian state. Their most pressing concern, therefore, is finding an effective means of erasing the gap between what is and what could be, between the real and the ideal.

It turns out, however, that the lesson to be drawn from "Jon" is not as simple as it might first seem. The ultimate irony of "Jon" is that it plays with the in/out duality only to subvert it. It shows this duality to be false metaphysics, in the final analysis. It is false because the product testing that goes on in the Facility is targeted for application beyond its walls. The "outside" world is organized in accordance with the information gathered from within the Facility. So, in a sense, Jon and Carolyn never fully exit the Facility. The outside world is as much a construct as the inside, although arguably a less totalistic one.

It follows, then, that the love Jon has for Carolyn cannot "save" him in any simple sense of the term because that love forever will have to contest with a social environment that challenges it. The only thing that has changed is how they perceive and relate to this order. We are led to believe at story's end that Jon and Carolyn now see it in perspective. The reality they thought was all encompassing and cool has been revealed to be smaller than once imagined and unrewarding. In short, what for them constitutes "the real" has changed, a reorientation that has the potential to change everything.

To conclude, "Jon" is a piece of speculative fiction that holds a mirror to our society by revealing the depth of our complicity in a consumer society that wants what technology more generally wants, the production of self-satisfied beings. Saunders appears to remain optimistic that our intractable human nature is a bulwark against the drive to self-satisfaction. Saunders says as much himself. Reflecting on "Jon," he notes: "Despite the world he [Jon] lives in, his emotion is not stunted, although his language is. He feels, but his lens, his speaker is too small, and it doesn't mean he's not feeling. That experience, for me, is somehow where we are as a culture." But, after this relatively encouraging assessment, Saunders immediately pulls back, saying, "That is the frightening thing to me—that the inability to express yourself results ultimately in the inability to feel." Saunders admits here that feeling and expression are not divisible, that context matters. Emotion—and an emotion as primal as love—*can* be stunted if the conditions in place are hostile to its cultivation and refinement.[46]

So love may not remain unconquerable after all. Saunders seems to suggest that we deceive ourselves if we think it impervious to assault. Surely, this much can be gleaned from "Jon." *Eros* is corruptible: our erotic attachment to the world is corruptible. It is not predetermined that we experience and reflect on *ta erotika*. If "Jon" is a love story, it is just barely. And if "Jon" mirrors our world, then Saunders alerts us to the impact of the deeroticizing of our social order. Less and less do we see the world through Socratic eyes as a place where we participate in an open-ended order of being. Our horizon, instead, is shrinking. We live increasingly in a realm of our own making, divorced from any sensibility that might transcend its own internal operational logic. This diminution amounts to a loss of perspective. As argued, the one thing Socrates understood with clarity was the extent to which he was moved by the world around him. Mind for him is not self-contained but exists in erotic tension with the natural order, in both its intramundane and cosmic dimensions. In contrast, Jon's world—and our world, by extension—is almost entirely self-enclosed.

Closing the gap between desire and its satisfaction is modernity's life-work. We are encouraged to want what we want and to get it, or to get at least some semblance of what it is we desire so as not to dash the illusion of entitlement. This psychical ideal is antithetical to a Socratic understanding, which, to repeat, focuses precisely on what our desiring tells us about our placement within the world. Saunders informs us that we live in a world that sees the erotic soul as counterproductive to its own functioning. Business school gurus like to tell us that disruption is good. But, of course, the kind of disruption they have in mind is the kind that leads to innovation *within* the existing operational system, the kind that further enhances efficiencies and profitability. To this extent, creative disruption further entrenches the Facility-like character of the contemporary social order. It is a consolidating strategy for problem solving, not a strategy for questioning the problem-solving ethos that pervades the technological order.

Eros is a truly disruptive force. Socrates's disquietude led to this questioning the conventional wisdoms of his day. In the process a space opened between received opinion and wisdom, the very space of philosophical pursuit. Jon's erotic attraction to Caroline likewise precipitated a shift in perspective that enabled him see beyond the parameters of right understanding as established within his social order.

NOTES

1. As recited in Plato's *Theaetetus*, Protagoras is reported to have said, "Man is the measure of all things—alike of the being of things that are and of the not-being of things that are not." (152a)

2. This claim is made in reference to Socrates's statement: "You must forgive me, dear friend; I'm a lover of learning, and trees and the open country won't teach me anything whereas men in the town do." Plato, "Phaedrus," in *The Collected Dialogues of Plato*, ed. Edith Hamilton (Princeton, NJ: Princeton University Press, 1961), 230d.

3. See Plato's *Apology*, 20d.

4. Aristophanes's spoof of Socrates's natural philosophy is best exemplified in the Parodos section of *The Clouds*.

5. *Apology*, 19c.

6. Plato, *Phaedrus*, 227a.

7. Ibid., 228d.

8. Ibid., 230c–d.

9. Ibid., 236e.

10. Ibid., 237d.

11. Ibid., 237d.

12. Ibid., 237d.

13. Ibid., 238c.

14. Ibid., 238d.

15. Ibid., 244b.

16. Ibid., 245c.

17. Ibid., 246a–254d.

18. Plato, *Timaeus*, 36e–37c; 47e–48a.

19. The theme of disenchantment is addressed more thoroughly in chapter 2.

20. Thomas Nagel, *Mind and Cosmos: Why the Materialist Neo-Darwinian Conception of Nature Is Almost Certainly False* (Oxford: Oxford University Press, 2012), 123.

21. Plato, *Symposium*, 178d.

22. Plato, *Laws*, 644e.

23. An earlier version of my analysis of George Saunders's "Jon" first appeared in *VoegelinView*, an interdisciplinary and international website dedicated to the thought of Eric Voegelin. The original article, entitled "It's Not Yours To Do With What You Like!: A Critical Reading of George Saunders's 'Jon,'" can be retrieved at: https://voegelinview.com/like-critical-reading-george-saunders-jon/

24. George Saunders's short stories have been winning national and international literary awards since the mid-1990s; he won the MacArthur Fellowship in 2006. His latest collection of stories, *Tenth of December: Stories*, has been on numerous bestseller lists since its publication in 2013. "Jon" was first published in the January 27, 2003 edition of *The New Yorker*, and has been subsequently republished in the collection, *In Persuasion Nation* (New York: Riverhead Books, 2006).

25. Total Quality Management can be defined as a business approach that focuses on performance feedback and its progressive refinement as a chief means of improving the quality of products and services. The effective gathering and assessment of data is central to TQM, thus fulfilling John von Neumann's 1949 prophecy that in the future the main technological challenge will pertain to concerns over "organization, information, and control."

26. George Saunders, *In Persuasion Nation* (New York: Riverhead Books, 2006), 23.

27. Ibid., 23.

28. Ibid., 24.

29. Ibid., 26.

30. *Exodus*, 33:3

31. See Paul Feyerabend's *Conquest of Abundance: A Tale of Abstraction Versus the Richness of Being*, ed. Bert Terpstra (Chicago: The University of Chicago Press, 2001).

32. *In Persuasion Nation*, 27.

33. Ibid., 28.

34. Ibid., 29.

35. Ibid., 29.

36. Ibid., 30.

37. Ibid., 30.

38. Ibid., 47.

39. Ibid.

40. As Winston opines, "The future belonged to the proles." George Orwell, *Nineteen Eighty-Four* (London: Penguin, 1989), 229.

41. *In Persuasion Nation*, 55.

42. Ibid., 55.

43. The passage from *Nineteen Eighty-Four* reads, "As he looked at the woman in her characteristic attitude, her thick arms reaching up for the line, her powerful mare-like buttocks protruded, it struck him for the first time that she was beautiful. It had never before occurred to him that the body of the woman of fifty, blown up to monstrous dimensions by childbearing, then hardened, roughened by work till it was coarse in the grain like an over-ripe turnip, could be beautiful." Ibid., 228.

44. *Periagoge* means a "turning around" or conversion toward a philosophical life, or a life in pursuit of the true and the good. As related in the *Republic* (518c–d), the turn toward a love of wisdom is a matter of redirecting the soul's vision, not implanting insight into the soul.

45. I am thinking here of the statement attributed to Horace: "*Naturam expellas furca, tamen usque recurret,*" or, "You may drive nature out with a pitchfork, but she will always come back."

46. In 2008, "Jon" was first staged by the Chicago-based theater company, Collaboraction. The quotations cited in this paragraph are drawn from a conversation George Saunders had

with Monica Weston, at *Newcity Stage*, and is available at:http://www.newcitystage.com/2008/
10/27/consumed-with-desire/.

Chapter Two

The Ethos of Technology

Highlighting the erotic dimension of human experience is a prelude to thinking about technology. Ultimately, thinking about technology is an exercise in reflecting on *eros*. Socrates understood well that the human mystery is rooted in the perception of this world. The world of common experience, by virtue of its perceived orderliness, suggested to him the existence of a more encompassing cosmic order whose nature could be deduced only indirectly through its representations. To be fully human meant for him to remain attentive to the extent to which the human drama plays itself out against the backdrop of a more expansive canvas. I am positing here that being human for us means something quite distinct from what it meant for the ancient pagan philosopher. But before we are in a position to elaborate upon this distinction, the next stop on the path to thinking about technology brings us to the question about reality and its perception.

American novelist and essayist Marilynne Robinson articulates what arguably is an underappreciated truism. "We inhabit, we are part of," she says, "a reality for which explanation is much too poor and small."[1] By "explanation," it seems, Robinson means any rational account of the whole, be it secular or religious. No possible account, in her estimation, is adequate to the task of supplying a coherent and total explanation of the order and meaning of things. Explanation or interpretation is necessarily incomplete, in other words. Reality outflanks our interpretive capacities: it is always "more" than we take it to be. Now, Robinson's philosophical skepticism serves her religious purposes well. Should reality remains at bottom a mystery, it is rather fortunate that the kind of discourse traditionally most open to the irreducible unfathomability of reality is religious discourse, or at least certain strains of it. It is important, however, to note that this skeptical orientation is capable of standing on its own. Unaided reason can inform us that the real world of

21

everyday experience does not submit to any totalistic forms of explanation. In other words, an argument can be made in support of the conclusion that there are limits to rational understanding.

Let us begin to defend the proposition concerning reality's unfathomability by recounting Niklas Luhmann's koan-like observation: "The world is observable *because* it is unobservable."[2] The gist of Luhmann's argument can be explained relatively simply, and perhaps most simply by taking the observability of the world in its literal, visual, sense. Seeing by definition is a delimiting act. To be able to perceive anything requires our not being able to perceive everything, which is akin to saying that perception is perspectival. We necessarily see from *somewhere*. The world always opens up before us from some vantage point, which precludes the possibility of total vision or the absolute gaze. To observe, then, is to have the world divided into the seen and unseen. The condition of possibility for vision is invisibility. Differently put, we might say that the price paid for the capacity to observe the world is that the world in its totality remains hidden from view.

The same phenomenological restrictions apply within the realm of cognitive perception. Seeing with the mind's eye is a perspectival act. All ideas, including ideas pertaining to the nature of the whole—scientific or otherwise—are theoretical constructs, which, by their nature, are cut from the cloth of all possible ideas and therefore marked by their partiality, by theoretical blindness. A Theory of Everything is thus chimerical. Epistemologically, we are barred from entry into the domain of Truth. Built into the nature of things are the conditions that prohibit the part gaining access to the whole.

What follows from the above consideration? One important point to be drawn is the existence of an irreducible gap between reality and our conceptualizing it. There is, as it were, a space between the real and our understanding of it that militates against their conflation. There is something about the constitution of reality that withholds from us a clear and transparent picture of its constitution. In more evocative language, we might say that there is something surreal and uncanny about reality. The earthly arena within which the human drama has played itself out for millennia remains constant in a way its perception does not. We may inhabit one earth, but we have lived in many worlds or interpretive realities. Cognizance of the "many worlds" thesis is a key element in what it means to think about technology, the full import of which will be addressed later. For present purposes it is enough to note that phenomenal reality never simply "is" naked before us in its unconcealed truth. It is never present to us in its brute facticity because "the real world" is always already an imagined reality. That is to say, reality is from the start seen as meaningful, and what counts as meaningful in one context need not count as meaningful in another.

Some persons might bristle at the suggestion that reality is, in any sense, "illusory." I only can repeat in response that the brute facticity of the "real world" has never been questioned. All that has been said here is that because reality exists for us, because it means something to us, reality is necessarily theory or picture-dependent. And it is because there are and have been many picturings of the real that we can say there are many realities. It follows from this that how the world appears is determined by the dominant imagining of reality at a given time and place.

This pivotal observation is simple yet radical in its implications. As said, it means that reality can never be seen or understood as it "really" is. It means reality—the natural order of which we are a part—can never be known "objectively," as if the knower were positioned beyond the thing known, peering at it from the outside. There is, in short, no context-free understanding of the real. While the phenomenal world is real, or present to our senses, its meaning is never fixed or absolute. To the contrary, we read reality as we read a literary work and, like a good book, the world is capable of revealing more than one meaning. As a result, reality is what we make it, as determined by assumptions regarding its meaning and purpose, assumptions that are context dependent and therefore variable. To conclude, we can say that because meaning is refracted through human sensibility, meaning is always multiple. The world we are born into yields many realities or worldviews.

This insight is foundational to my argument and can hardly be overstated. It informs us that there is a gap, an irrevocable gap, between what the world offers us and what we glean from it. For all our efforts to pin it down, the meaning of the world cannot be resolutely fixed. It eludes us. It resists capture. To this extent, the world remains "other" than us or alien to our desire for settled meaning.

The time has come to align more closely our discussion of reality as an interpretive phenomenon with the subject of our investigation—technology. When we speak of technology today, and we talk about it incessantly, most of the discussion appears to fall into one of two categories, both of which focus on technology as an instrumental power and for that reason are marked by a certain unthinkingness. The first line of commentary assesses technology in terms of technological innovation by examining its capacity to render the world more responsive to perceived needs. These innovations typically are considered in accordance with their capacity to deliver on some promised good, like a cleaner environment, more effective communication, or improved health. Analyses of this type address technology in strictly technological terms. By this is meant that one strain of popular literature on technology judges technology according to standards internal to its own functioning. If technology, in broad terms, has as its end the betterment of humankind, then emerging technologies are assessed in relation to their promise to realize that end. Will, for example, new schoolroom technologies make for smarter students? Will wearable technology make for a healthier citizenry? These kinds of questions ask of our technological tools whether they work as intended. Two types of answers emerge from this line of inquiry. The preponderant response is affirmative. Yes, we are told, the sequencing of the human genome is ushering in a revolution in the way disease is treated. Or, yes, Big Data is in the process of saving politics from its own ineptitude.

To be fair, the boosterism that pervades much technology-related commentary often is leavened with a measure of caution. For a variety of reasons, some more sincere than others, warnings frequently accompany the promotion of emerging technologies, like those lists of counterindications at the end of pharmaceutical ads. But sometimes not. A subspecies of techno-boosterism is unashamedly bullish on the theme of technological advance. Self-proclaimed futurists such as Ray Kurzweil, Kevin Kelly, and Clay Shirky sing rhapsodic about the blessing to humanity that is technology, and like communists of old, prophesy the best is yet to come.[3] Then, as if to redress a cosmic imbalance, we witness an apocalyptic strain of literature that targets technology as the source of immanent civilizational collapse.[4] For all its gloom, this reading shares with its roseate counterpart an instrumental focus. We are informed that technology is so profoundly dysfunctional that it is precipitating an imminent return to the Stone Age.

Now, there is nothing inherently misguided about thinking about technology along technological lines. Whether or not technology is delivering the goods as advertised is a legitimate concern that merits serious inquiry. But too much is left out of the picture if technology is approached almost exclusively in these terms, as it appears to be, especially within the realm of public discourse. Indeed, the welter of informed dialogue about technology is so skewed in the direction of instrumental concerns (and has been for generations) that an argument can be made that we have all but lost sight of alternative ways of assessing technology.

We are locked into thinking about technology in instrumental terms in part because of its pervasiveness in our lives. We are least likely to pointedly reflect on what is closest to us, and nothing is closer to us today than technology. Like water is to fish, we swim in an ocean of instrumentalities so intrinsic to our functioning it has all but disappeared from view. Our oneness with technology works against efforts to see technology for what it is, to gain a "big picture" view of the phenomenon.

Why this long view of technology is important and necessary will be the focus of more sustained debate later. Our more pressing concern here is to abjure allegiance to the view that associates thinking about technology with assessing whether it is "good" or "bad." The important question is not whether technology works or not but how it has altered our view of the world and how we see our place within it. Exploring this question is an exercise in thinking about technology. Unlike what passes these days for thinking about technology, the line of reasoning pursued in these pages aims to examine technology from an extra-technological vantage point. The goal is not to think within the parameters of technological thought but to situate oneself in a position where it is possible to think *about* these parameters and, accordingly, to assess their limits and possibilities. To imaginatively reposition oneself in this way is to enter into an alternative world to the technological. This repositioning is necessary because to see technology for what it is requires seeing what is excluded from the picture of reality that informs the technological project. We have to stand back in order to see the broader canvas upon which the technological vision is projected.

I have just made a logical leap that demands clarification. By linking technology with a "picture of reality," I have said something important about technology that today remains largely unsaid, namely, that technology is as much about an idea as anything else. According to this reading, the most significant reason why technology eludes critical reflection has less to do with the ubiquity of technologies than with technology's status as an idea. Technology's hold on us is near total, and almost totally unseen, because the vision of the world that supports the technological enterprise is largely invisible to us. Technology rules, arguably, because its guiding assumptions regarding the nature of reality and our place within it comprise our everyday or commonsense understanding of the way things are—of reality. Having become desensitized to our civilizational bias toward the technological, we have lost sight of any way of perceiving the world other than the technological.

Against the common identification of technology with tool use, it is suggested here that thinking about technology requires that we think about technology as a worldview, or a grand idea relating to reality and its meaning. So what is the technological worldview? What constitutes technology's distinctive spirit or ethos? As already alluded to, the technological outlook perceives the world to be an object of potential or actual control, a resource to be used for the purported betterment of humankind. As a perception, technology is (to borrow from Heidegger) a revealing where the world shows itself to be an object of use and mastery.[5] The world's "reality" is manifested in its usefulness. In the age of technology, we gain our bearings in the world by asking, "What can be done with this?" In a world where everything is assumed to be a resource for human use, guiding questions are geared to determining the means and ends of resource deployment.

Technology in popular discourse is usually not thought of in this way, and for good reason. For once a worldview gets established, it disappears as a reading of reality and instead becomes identified with reality itself. The interpretation becomes the thing interpreted. History turns into nature when a time-bound, determinate interpretation of reality is identified with the unchanging truth of the real. As with the identification of opinion and truth, the perspectival character of perception is occluded with the consolidation of a perceptual framework. So it is that in the technological era the world simply "appears" as a resource to be managed, and technology likewise the "natural" means of resolving management related issues. Lost is an awareness that this pragmatic, commonsense orientation issues from a certain perceptual predilection to see the world primarily in utilitarian or use-related terms. Pragmatism, strangely enough, is an effect of a prior cause in the form of a particular understanding of the meaning of the real. We moderns are placed on Earth not to be heroes, as was self-evident to the Homeric Greeks, but to subdue the ground we stand on.

The power of worldviews resides in their invisibility. This also applies to the power of technology as a worldview. What goes undetected is that something must pertain *before* technology, as a set of instrumental practices, can come into being. This "something," to repeat, is a particular perception of the world, one that debuted in Europe roughly five hundred years ago.

The Renaissance marks the time period when Western civilization began its transition out of a worldview centered on a Christian reading of reality. This shift marked our entry into modernity, which is synonymous with the reinterpretation of reality along technological lines. One of the first and best illustrations of the modern or technological attitude in Western philosophical literature is contained in Niccolò Machiavelli's *The Prince*. What makes this sixteenth-century political treatise a distinctly modern work is the attitude it conveys toward fate or destiny. In the world Machiavelli was working to leave behind, there was a prevailing sense that humans were actors in a game governed by larger, transcendent forces that severely constricted the realm of what was considered humanly achievable. In this sense the Christian world resembled the ancient Greek: it was assumed there were forces at play other than human forces. Their power aside, political leaders were hamstrung by this sensibility. They, too, were resigned in their acceptance of the limits of political ingenuity.

Machiavelli challenged this notion. He wanted to convince aspiring autocrats that, with sufficient insight and effort, they were capable of mastering their political fortunes to an unprecedented degree.[6] He wished to curb the fatalism that gripped so many of his fellow Florentines, to alter perceptions as to what was possible in order to extend the art of control. Simply put, Machiavelli wanted to modernize the mind-set of the ruling class. And most importantly he understood that no teaching outlining the tricks of the political trade, as potent as this teaching might be, would be effective in practice without a change in spirit or outlook. An effective "make happen" leader assumes a corresponding make-happen attitude, the very attitude that today lends our age its technological character.

It is important to realize that there was nothing inevitable about the West's entry into modernity. Clearly, not all cultures followed the West's path. Some still have not modernized to this day. So the emergence of the technological worldview was not the result of some "natural" unfolding process. Rather, it was fought for, and texts like *The Prince* were pivotal in the struggle to change perceptions regarding what was possible within the realm of human action.

The scenario just described needs to be stated in more general terms before we can proceed with an accounting of technology as a world picture. The reformulated story goes something like this: Humans once assumed there existed permanent and insurmountable barriers to the understanding and mastery of the world, including ourselves as worldly beings. This assumption prevailed because the ways of the world were presumed guided by the gods or a god, whose designs for us, by virtue of their transcendence, exceeded the bounds of mere human comprehension. Perceived as grounded in the supernatural, the natural and human orders possessed an alien quality that rendered void any effort to make transparent sense of the world. So whatever steps may have been taken to fathom reality, to give a mythical or rational account of it, were played out against the backdrop of the ultimate inscrutability of reality. In a so-called enchanted world, not all things are knowable, not even in principle.[7]

The perception that the world was an intrinsically mysterious place conditioned our understanding of the world, and along with it, our actions. Assuming we humans were not privy to the ultimate workings of reality, we had no aspirations to extend indefinitely whatever powers of control we may have exerted over the world. History is seriously misread if it is assumed that premoderns were moderns-in-waiting, sidelined by frivolous cultural distractions that prevented them from getting on with the "true business" of life. No one, for example, could claim the ancient Greeks lacked the ingenuity to make a serious go of technological development. Any deficit in this regard was a reflection of their specific cultural bent toward what constituted the good life. So the Greeks were no proto-moderns. They were a tool-using people with an interest in the science of craftsmanship, or *techne*, as they called it.[8] But for all this they had no appreciation of technology as we moderns understand it. They did not because other kinds of rational pursuit (i.e., theology, ethics, politics, mathematics) were considered more worthy of serious attention. Not averse to the technical arts, the Greeks just weren't into them.

This instructive insight about the Hellenes and other premodern peoples is largely lost on us today. We are given to think their relative indifference to the technical arts was a reflection of a vexing incuriosity toward nature's workings and securing the means of human betterment. But incurious they were not. They merely understood the world in a way that drew from them a set of responses to the question of meaning and purpose at odds with our own.

The specifically modern, technological response to the question of meaning and purpose is grounded in the assumption that the world is disenchanted, or devoid of inherent meaning, and that meaning necessarily issues from we humans and us alone. So if something is perceived as valuable or beautiful or meaningful, we believe it is because we confer these qualities upon it, not because the admired thing is thought to possess any intrinsic worth along those lines. It follows that a disenchanted world is a mute world: we speak for it given the assumption that it cannot speak for itself.

To the extent the spirit of enchantment lives on today it resides primarily within certain indigenous cultures. While we moderns may respect this premodern worldview, perhaps even envy those for whom it still holds a measure of legitimacy, we realize at the same time that this vision is not our own. It does not reflect our understanding of the world we inhabit. The modern understanding rejects the notion that we are cut from the same cloth as the world, that we are participants in a reality open to the whole of being and therefore live erotically charged existences. Lost is the sense that human and world are ontologically bound. Instead, we live in Descartes's shadow, predisposed to see ourselves pitted against *res extensa* (extended matter), the inert "stuff" of the material universe that slavishly obeys preset laws.[9]

Possessing what is presumed that nonhuman nature lacks, we take upon ourselves the task of enlivening the world by reordering it, by humanizing the inhuman. In speaking for the world and bestowing meaning upon it, we see ourselves as redeeming an otherwise derelict and alien habitat. The storehouse that is nature is saved from itself—from its blind fatefulness—by being put to human use. All of this to say that the phenomenon called technology issues from a deep existential unease with the world into which we are born.

What Friedrich Nietzsche asked of morality in *On the Genealogy of Morality* we must ask with respect to technology: What is the *value* of technology? What does our commitment to the technological project reveal about we hold to be important and true? What is the itch technology scratches, the repair it effects? We cannot begin to understand technology and assess its impact without first exploring what technology means to us given the cultural context within which it emerged.

The suggestion here is that thinking about technology requires we first abandon the notion that we moderns are merely sophisticated tool users. We misread matters if we assume that, like other civilizations before us, we are driven by a universal interest in easing the burdens of the human estate that we just happen to be particularly adept today at capitalizing upon. It is misleading to think we are better at playing the same game as our forebears. Our world and theirs are incommensurable. They are categorically distinct because tool use for us is grounded in an ethos distinct from theirs, a distinction that changes everything.

The difference can be stated succinctly. Only in a technological society do the practical arts serve as a refuge from the doubt that sustains the zetetic or inquiring spirit. Only in a technological society does the project to relieve the human estate assume sufficient proportions that questions pertaining to the good life find no other expression than through a technological frame. Technology's "gift" is freedom from the burden of thinking. Ancient Athens had its triremes and clepsydrae[10] but also its sibyls and Socrates. Its spirit of technological inventiveness was bound by a cultural setting that remained open to questions concerning the good life. Instrumental concerns therefore were never entirely divorced from concerns over ends. In short, sight was never lost of the larger question regarding the *value* of the practical arts. They remained for the ancients a means to an end, a tool employed for a purpose other than mere tool use. Arguably, the reverse is the case today. A technological society is a society for which the deployment of the practical arts becomes an end in itself. In this regard, technology today serves the same historical function as organized religion once did. It supplies both a vision of the good life and the means of its achievement.

It is therefore too mundane a valuation of technology to consider it merely in instrumental terms. Technology instead holds ideological power.[11] It provides us with a set of ideas that helps make coherent sense of the world and our place within it. Technology is, in short, a system of thought that, like all systems, highlights certain features of reality to the exclusion of others, which is to say it presents us with a settled "picture" of reality. Like its more straightforwardly political cousins, the image of reality technology advances aims to simplify the complex, to resolve an otherwise indeterminate reality. This ideological function is realized in several ways. First, as noted, the technological ethos helps supply the world with meaning and ourselves with purpose. Against Aristotle's *zoon logikon* stands our *homo faber*.[12] There is no better window into the soul of "man the maker" than John Locke's observation that the earth has been given to us for our "support and comfort," and that we humans are sanctioned to "make use of it to the best advantage of life, and convenience."[13] For we moderns, "making use" is hallowed business.

Another ideological function of technology is evaluative. The technological mind-set carries with it a sense of propriety. Certain ways of acting are deemed more approbative than others, given technology's explanatory bias. There are, in other words, positive and negative values associated with the technological outlook. One can evaluate what constitutes progress in a technological society by determining to what degree a particular development advances a value aligned with the technological ideal, such as efficiency for instance.

Lastly, technology as ideology provides a programmatic function. It offers us a way forward, a guide for future development. Since technology by definition is an action-oriented belief system, its ideology is its program. To live in a technological society is to know one's role as a foot soldier in the Grand March of progress. The path forward is set: tomorrow will look much like today with the exception that we will have taken another step in extending our powers of mastery.

Technology assumes the function of an ideology because it constitutes the overarching interpretive framework within which our societal order unfolds. We see the world through a technological lens: everything is refracted through the prism of its ethos. That is why technology rightfully can be called modernity's independent variable.[14] It is because we see the world in technological terms that the spirit of technology affects outcomes in all domains of societal life. In short, technology leads. It is not an effect of any cause external to its internal dynamic. Consider, as a case in point, the expression "media shock."[15] Media shock refers to the impact recent technological advances have had on traditional media institutions and the political landscape. Twitter is one component of the new media shockwave with well-documented effects on both journalistic and political practices and outcomes. The reverberations emanating from this and other like innovations are unidirectional. Technological change induces societal adaptation, not the reverse. We are, in other words, unlikely to witness a "politics shock" whereby technological development is shaped by policy means in a direction that subsumes its advance to ends external to its own immanent logic. While this causal reversal might constitute the ideal for someone like Jürgen Habermas,[16] who argues in favor of restraining technological rationality through broad-based political means, there is scant evidence to suggest that the technological genie is about to be put back into its bottle. Doing so would require breaking the ideological spell cast by technology, and signal our entry into a post-technological order, for which there few signs.

Clear thinking, in fact, points in the opposite direction. There is a veritable avalanche of support for the view that we remain in the thrall of the technological ethos. Take as evidence a recent advertisement for a major telecommunications company entitled, "When You Believe More, You Sleep Less." It reads: "Passion. Smart people working overtime. Running on fumes. Because they know what technology can solve."[17] The ad suggests that "smart" people know or believe (the ambiguity is revealing) what many fail to appreciate, namely, the limitless power of technology to solve problems. It is the cognoscenti's awareness of technology's boundless power, along with knowledge of their leading role in developing this power, that supplies them with the zeal of erstwhile missionaries. Like possessed souls, they eschew sleep and run on "fumes" while working feverishly to realize the endless potential embedded in the forces of good called technology.

This giddy rhetoric, typical of the boosters of technology (and driven by the profit motive), bathes technology in a distinctly religious light. Technology is the way to Truth, and only through technology can the Truth be revealed. The Christian promise to make right an unredeemed world is answered in the rhetoric of technology's boundless potential to solve the world's problems, in the here and now. We have convinced ourselves we are fighting the good fight when wits are sharpened and resources deployed to revitalize a decadent world.

It is axiomatic that no right-minded person is at peace with the world's iniquities. Total acquiescence in the face of life's injustices is neither possible nor desirable. Yet the offense taken today at the world's sufferings and lesser inconveniences reveals a profound and pathological discomfort with the world as it has been given to us. With modernity, life has been refigured as a solvable problem. Ours aspires to be a tragedy-free age. The world as it is must make way for the world as it should be. "Heaven on Earth" is modernity's unofficial motto. The belief—supplied by the technological ethos—that the gap between the real and the ideal is resolvable defines us. It also is the source of our heightened anxiety. In expectation of the assurances technology promises, we lose the resolve needed to cope with life's unavoidable pains and uncertainties, a loss that in turn exacerbates our anxiety about the human predicament and makes us more reliant still on finding technological fixes for our perceived ills.

The extent of our unease today is revealed by our total commitment to the technological project. In the words of Albert Camus, "we extol one thing and one thing alone: a future world in which reason will reign supreme."[18] Our commitment to reason, to scientific reason, and to the project of extending indefinitely our powers of mastery over nature, is near absolute. We have convinced ourselves that everything must be brought under the yoke of reason and managed according to scientific, rational principles. This requires that everything be operationalized or conceived as a "system" to enhance the powers of efficacy. Nothing is spared systematizing. From management practices to toothbrushes, from transportation networks to sneakers, everything is being reconceived as an assemblage of parts configured in ways to maximize performance.

For all our productivity, we moderns lack balance. We have been captured by the technological ethic and its singular commitment to the ideology of domination and control. This explains, as noted, our growing (and irrational) aversion to suffering. In a culture of control, nothing is more unpalatable than having to endure an experience over which we have no control or input. Everything, to the contrary, must be open to manipulation. Consider, in this regard, our societal response to gender and death, those traditional markers of given nature.

So everyone is a Machiavellian today. We have rejected utterly the notion that life is something to be accepted on its own terms. "Fate" is a dirty word. The objective of existence, we tell ourselves, is to create and refine the means of determining our own life path, both individually and collectively. And this is precisely what we set out to do. Centuries ago we reconceived the world as radically reformable, laying the ground for the birth of modern science and the technological project. We since have pursued the domestication of the world in an effort to create something resembling an Earthly paradise.

It must be reiterated that the dominion of the earth via technology has little do with mere tool use. Humans are tool-using animals and have been for a good portion of the species' long history. What makes our age the age of technology is the value placed on this specific capacity of ours. As observed, in premodern times the value of the technical arts was deemed subordinate to other, higher kinds of goods. In Homeric Greece, for instance, the noblest life was identified with the life of the military hero and the virtues associated with this life, such as courage and loyalty. In medieval Europe the saint replaced the hero as the highest human type, again assigning to the productive arts an inferior position in the rank ordering of goods. Only in the modern world do the technical arts attain an independent dignity. No longer in the service of a higher end or good, the decontextualized technical arts are valued in themselves as worthy of cultivation. "How" questions come to dominate the public agenda, displacing older modes of inquiry and conduct that focused on the ends of human existence, or on questions of meaning and purpose. As the technical arts were liberated from ends transcending the realm of the technical proper, the instrumental ethic came to be regarded as an intrinsic good. Seen as a power in its own right, technology's end became identified with its own unfolding or self-development.

But what kind of unfolding are we speaking about here? What internal logic specific to the technological enterprise accounts for its progression? To answer this question we have to revisit the claim that technology, as a worldview, takes the natural order to be a manipulable resource. In this context, the logic of technology (understood as an instrumental power) is revealed as the progressive unfolding of the powers of manipulation and control. This extending takes two broad forms. The first and most obvious form relates to the increasing reach of technological control. Technology progresses as new domains are brought under the ambit of "the technological." Progress here means the "resource-izing" of the raw stuff of nature, or the extension of our capacity to act into nature. The human gene, for example, has been turned into a manipulable resource as a result of recent scientific advance. Genes have been brought into the realm of the human. They have become patentable property and objects of therapeutic intervention. But apart from lateral extensions of this kind, powers of control also can be extended through their tightening. Here is where the issue of efficiency enters into the picture.

To illustrate what is meant by efficiency and its relevance to technology, consider for a moment the field of communications. As often told, from the time of antiquity to roughly 150 years ago, the fastest a message could be transmitted across land was on horseback. The pace of a galloping steed was the communications speed limit for millennia. In 1860, for instance, it took a little over seven days for Californians to read about Abraham Lincoln's election as president, and that was record time. Since then, several waves of technological advance have brought us to the age of electronic messaging, where the traditional constraints associated with distance communication all but have been obliterated.

Why is it that the Pony Express predated instant messaging and not the reverse? Why is the quill the ancestor of the ballpoint pen and not its descendant? What is the spark that animates technological evolution and lends it force and direction? In a word, efficiency. Technological advance is measured by gains in efficiency or the economical use of resources consumed in the realizing of a targeted goal. Differently put, efficiency refers to the ratio between inputs and outputs, between the resources expended in performing a technical feat and the productive capacity of that performance. This ratio can be maximized in one of three ways: first, if outputs are held steady, by lessening inputs; second, if inputs are held, by increasing outputs; or lastly, by simultaneously increasing yields and reducing the expenditure of resources needed to produce these yields, the resources in question including time, money, and materiel.

So the Pony Express, like other defunct practices, has been abandoned because human ingenuity has found more economical ways of performing given tasks. A technological society, then, is defined not just in terms of control but equally in terms of efficient control. Just as capitalism denotes not simply a wealth-producing economic system but a system that finds in wealth production the means for greater wealth production, tool use in a technological society is driven by a logic that seeks to create ever more efficient means of tool use. Progress for us is measured by how closely a technology approaches perfect efficiency. It is self-evident in such a context that the "right" way to do things is the most efficient way. It is axiomatic as well that the commitment to efficiency is ongoing. It stands to reason that if technology has become an autonomous enterprise, then efficiency, its motor, is always running. Like Friedrich Nietzsche's understanding of the will to power, the efficiency business is the business of *increasing* efficiencies. [19] Its pursuit is relentless. More always can be done to reduce waste or increase productivity in the realization of some end. Technologies are deployed in succession in the hunt for the unattainable ideal of perfect efficiency or perfect functionality. It does not matter whether we are talking about the evolution of the microchip or inventory systems or airline design, the common thread of all technological development is the chase for greater efficiencies. So modern technology is never simply about expanding the range of our powers of control over nature but equally about extending the powers of efficient control. The one aspect cannot be thought without the other.

If, as I argue here, efficiency is a central feature of technology and a primary value in a technological society, then efficiency trumps all other goods or values. Living in the world we do, we are conditioned to interpret "progress" to mean advancements in efficiency. This advancement is the measure of our moving forward. And by pointing the way ahead, the drive to increase levels of efficient control lends meaning to the technological project and to our lives as contributors to this project.

NOTES

1. This quotation is extracted from a 2002 online essay Marilynne Robinson published with *The Chronicle of Higher Education*, entitled "Reclaiming a Sense of the Sacred." The essay can be found at:http://chronicle.com/article/Reclaiming-a-Sense-of-the/130705/.

2. Niklas Luhmann, "The Paradox of Observing Systems," *Cultural Critique*, no. 31 (Autumn, 1995), 46. The article can be retrieved online at: http://www.jstor.org/stable/1354444?seq=1#page_scan_tab_contents.

3. There is a long and ever-growing list of techno-utopian texts geared toward a general reading audience. Some of the more popular published with the past five years include: Ray Kurzweil's *How To Create A Mind: The Secret of Human Thought Revealed*, Peter Diamandis and Steven Kotler's *Abundance: The Future Is Brighter Than You Think*, Clay Shirky's *Cognitive Surplus: How Technology Makes Consumers Into Collaborators*, Daniel Kellmereit's *The Silent Intelligence: The Internet of Things*, Kevin Kelly's *The Inevitable: Understanding the 12 Technological Forces That Will Shape Our Future*, Alex Pentland's *Social Physics: How Social Networks Can Make Us Smarter*, Steven Kotler's *Tomorrowland: Our Journey From Science Fiction to Science Fact*, and Eric Brynjolfsson and Andrew McAfee's *The Second Machine Age: Work, Progress, and Prosperity in a Time of Brilliant Technologies*.

4. The most powerful techno-dystopian writings tend to belong to the realm of fiction. The modern tradition arguably begins in the early 1920s with Yevgeny Zamyatin's *We*. Most recent contributions include Cormac McCarthy's *The Road*, Kazuo Ishiguro's *Never Let Me Go*, Paolo Bacigalupi's *The Windup Girl*, Michel Houellebecq's *The Possibility of an Island*, and David Mitchell's *Cloud Atlas*.

5. As Martin Heidegger says, "Instrumentality is considered to be the fundamental characteristic of technology. If we inquire, step by step, into what technology, represented as means, actually is, then we shall arrive at revealing. The possibility of all productive manufacturing lies in revealing." *The Question Concerning Technology and Other Essays*, tr. Willian Lovitt (New York: Garland Publishing, 1977), 12.

6. The classic statement regarding the unleashing of political ingenuity is contained in chapter 15 of *The Prince*, where Machiavelli informs his readers that the key to effective political action lies in distinguishing "how things are in real life" from how they are discussed in "imaginary republics." Politics is a dirty game, he notes, and leadership that does not take this elemental fact into consideration will quickly come to ruin.

7. Max Weber says a disenchanted world denotes not a physical state of affairs but an attitudinal orientation. The world, he says, is disenchanted with "the knowledge or belief that if one but wished one could learn it at any time. Hence, it means that principally there are no mysterious incalculable forces that come into play, but rather that one can, in principle, master all things by calculation." See, "Science as a Vocation," in *From Max Weber: Essays in Sociology*, tr. Hans Gerth and C. Wright Mills (New York: Oxford University Press, 1946), 139.

8. Aristotle says of *techne*: "Now, building is an art or applied science [*techne*], and it is essentially a characteristic or trained ability of rationally producing." *Techne*, therefore, is the

rational science or art of crafting into existence something that necessarily need not be. So *techne* pertains to the novel, not the natural, which for the ancient Greeks was precisely what must be. See Aristotle's *Nicomachean Ethics*, Book Six, 4.

9. In reference to the "two highest kinds of things," René Descartes says "the first of intellectual things, or such as have the power of thinking, including mind or thinking substance and its properties; the second, of material things, embracing extended substance, or body and its properties." See his *Principles of Philosophy*, 48.

10. A clepsydra is an ancient Greek water clock.

11. My framing a discussion of technology in ideological terms is indebted in part to Terence Ball and Richard Dagger's *Ideals and Ideologies: A Reader*, (New York: Longman, 2013). See, especially, "The Concept of Ideology," 1–2.

12. The term "homo faber" was popularized Max Frisch in his novel *Homo Faber: A Report*, tr. Michael Bullock (Fort Washington, PA: Harvest Books, 1994).

13. John Locke, *Second Treatise of Government*, ed. C. B. Macpherson (Indianapolis: Hackett, 1980), 26.

14. The claim that technology is the independent variable of the modern age is taken from Tom Darby's "On Globalization, Technology, and the New Justice," in *Globalization, Technology, and Philosophy*, eds. David Tabachnik and Toivo Koivukoski (Albany: State University of New York Press, 2004), 70.

15. The term "media shock" is given extended treatment in David Taras' *Digital Mosaic: Media, Power, and Identity in Canada* (Toronto: University of Toronto Press, 2015). See, especially, 7–26.

16. Of long-standing concern to Jürgen Habermas is regenerating the means to guide instrumental powers to properly human ends, as articulated through open and unforced public dialogue. See "Technology and Science as 'Ideology,'" in *Toward a Rational Society: Student Protest, Science, and Politics* (Boston: Beacon Press, 1970), 81–122. Also see "The Critique of Instrumental Reason," in *The Theory of Communicative Action*, vol.1, tr. Thomas McCarthy (Boston: Beacon Press, 1984), 366–99.

17. This Verizon advertising campaign ran in 2013.

18. Albert Camus, "Helen's Exile," in *Lyrical and Critical Essays*, ed. Philip Thody, tr. Ellen Conroy Kennedy (New York: Vintage Books, 1968), 149.

19. "My idea is that every specific body strives to become master over all space and to extend its force (—its will to power:) and to thrust back all that resists its extension. But it continually encounters similar efforts on the part of other bodies and ends by coming to an arrangement ("union") with those of them that are sufficiently related to it: thus they then conspire together for power. And the process goes on—" Friedrich Nietzsche, *The Will to Power*, tr. Walter Kaufmann and R. J. Hollingdale (New York: Vintage, 1968), s. 636.

Chapter Three

The Problem with Technology

Having examined the ethos of technology, we are in a position to probe further into what is problematic about its characteristic spirit. As mentioned in the introduction, one troublesome aspect is the expanse and grip of the technological vision. Technology is a global phenomenon. Unlike any world-view before it, technology is truly a planetary world picture. It seeks its completion in the phenomenon called globalization, whose end is the establishment of a global civilization given over to the project of technological domination. Not unexpectedly, the vanguard of technology welcomes the advent of "one world" and works toward its realization. Most assume its inevitability and so are largely indifferent to the phenomenon. They should not be, for reasons to be discussed later.

But the problem with technology is as deep as it is wide. That is to say, there is something about the technological vision itself that should give us pause for concern. This "something" is the imbalance that lies at the center of the technological outlook. It was suggested previously that virtually all worldviews prior to the technological world picture sought to lend meaning to the human experience in the context of the immutable order of things. In contrast, the technological vision challenges this order to yield to the human will. Acting on this challenge has resulted in our acquiring God-like powers over nature and over ourselves as natural beings; so much power, in fact, that the concept "nature" has been rendered all but redundant today.

Nature is passé because we moderns no longer subscribe to what it symbolizes. Since antiquity nature was thought equivalent to "the course of things," the unchanging and unchangeable ways of the world. By definition, the natural was what must be. It was against nature, thus defined, that the human drama played itself out. The "fate" of the world was juxtaposed against human efforts to effect change within the world. As we have seen,

37

Machiavelli, an early modern, sought to wrestle from fate a larger measure of human self-assertion than previously was thought possible. He settled for a fifty-fifty split.[1] In contrast, we late moderns want it all. As a matter of principle, our technological culture refuses to concede that there are any permanent obstacles to the attainment of total mastery. Problems exist to be solved, challenges to be overcome, in an endless struggle of human self-determination. So the technological project can be seen as a war effort seeking to vanquish the enemy "other" that frustrates the realizing of our dreams of omnipotence. This is a good fight, in our view. We are doing good work by refusing to accept there may be limits imposed upon us as natural beings in a natural world. This existential revolt against nature marks us as moderns. And our rejection of nature cannot be thought apart from our embrace of history, the arena within which we forge ahead with the project of mastery.

It is not surprising, then, given the "divine" powers of technology, that technology today is addressed habitually in reverential terms. The thinking about technology that goes on in the pages of *Scientific American*, the *MIT Review*, and scores of other more popular versions of the same, largely presupposes the sanctity of technology. Technology is consecrated before a word about it is uttered or a thought disseminated. With the publication of every successive "Ten New Breakthrough Technologies" list the doctrinal belief that technology holds salvific power is recapitulated. There is no space within this preset narrative to think critically about technology as a whole, to assess its impact in ways not already predetermined as worthy of analysis by the lights of the technological mindset itself. Why is this? Well, as just noted, one hardly would expect tough talk about technology to emerge from a culture that holds it with the kind of esteem once reserved for a deity. And there are economic interests to consider as well. It makes good business sense to valorize the force linked to the powers of productivity, efficiency, and functionality, and no sense not to. But there is a difference between promoting such a force—the power of technology—and promulgating it with the kind of blind, fundamentalist fervor we witness today.

Pecuniary interests alone cannot explain our love affair with technology. Our devotion to the cause cannot be reduced to practical considerations only, as important as they may be. Otherwise, how can you explain why it is easier to imagine the end of life as we know it than a world not given over to the endless pursuit of technological mastery? We cannot imagine a world beyond our own because technology for us is more a belief system than a mere practical strategy for relieving the human estate. Much like the transition of science into scientism, the practical arts of old have acquired an aura that has effectively anointed the "making" enterprise. If scientism is what science becomes when it is interpreted as the universal and uncontested path to true understanding, then technologism represents technology's fulfillment as an ideology. It is what happens to the transformative power of the productive arts when it alone becomes identified with the means of attaining the good life.

The scientific community is predisposed toward scientism, as might be expected. It has been for a long time. As the eighteenth-century social reformer Henri de Saint-Simon put it, "[I]t is because science provides the means to predict that it is useful, and that scientists are superior to all other men."[2] Saint-Simon's claim that the scientist is superior to all other human types is couched in universal terms. It is not accompanied with a caveat restricting the application of the term "human" to a specific people or time period. Not much has changed in this regard over the centuries. Contemporary physicist and author Lawrence Krauss is equally comfortable making the grand claim that science alone really matters. In his words, "At some level there might be ultimate questions that we [physicists] can't answer, but if we can answer the 'How?' questions, we should, because these are the questions that matter."[3]

Now, if the modern scientific approach alone is valid and science is inextricably aligned with the task of relieving the human estate, then the technoscientific (or technological) project of mastery likewise is authoritative. Technology rules the planet. But does it? Well, as with so many other issues, it depends on whom you ask. On the one hand, scientists in the mold of Richard Dawkins and Lawrence Krauss are genuinely peeved that scientific reasoning still meets with pockets of resistance in the form of religious belief, philosophical musing, and their cognates. The world is not nearly scientific and technological enough for them and for their sympathizers. On the other end of the spectrum we find a cohort that feels besieged by the advancing ranks of "specialists without spirit," those technicians whose vision is limited to the business of efficient management.[4]

Both the Krausses of this world and their nemeses agree that ours is primarily a technoscientific age. What they disagree over is what to make of this sociological fact. Is the ascendancy of the technological account of reality to be applauded or decried? To answer this question we have to consider what a world thoroughly infused by the technological ethos would resemble and compare it to a possible alternative.

To do so, we first must get into the heads of those who most fervently believe in the power of technology to deliver humanity from the evils of an unreconstructed reality. What does a planet wholly given over to the technological dream look like to this person of faith and to what end is technology's full realization a means? In general terms and in answer to the first part of the question, it would be a world uncluttered by obstacles to the forward thrust of technological mastery. Author Kevin Kelly, for one, has sketched such a world and presents a case in support of a global technological monoculture. His vision is instructive.

The planetary culture Kelly foresees developing is called the "technium." The neologism, coined in Kelly's *What Technology Wants*, refers not only to the "global, massively interconnected system of technology," but also to the societal superstructure that supports this system, including "the generative impulses of our inventions to encourage more tool making."[5] The story Kelly concocts in defense of the emerging technium is wildly speculative. It starts at the literal beginning, with the Big Bang, and ends in some unspecified future time with self-aware minds thinking up more minds that coalesce into a collective mind whose destiny is to "expand imagination in all directions" until it "reflects the infinite."[6] The certainty with which Kelly outlines the dialectical unfolding of cosmic and human history would make G. W. F. Hegel blush, but there you have it. Modern technology not only fits into the evolution of the cosmos as tidily as it dovetails with our own as a species, but it constitutes the path to the fulfillment of our species' cosmic destiny.

The notion that our seeing "the infinite" is conditional on getting out of the way of humanity's co-evolution with technology is a bad advertisement for the cause. Apart from its ridiculous hubris, the assertion that our progress as a species is best served by disappearing into our creations is profoundly anti-humanistic and deeply disturbing for that reason. There is a lot of conspicuous self-loathing going on in much of the technology-will-save-us literature. The genre's propensity is to portray humanity in Nietzschean terms as a rope over an abyss connecting what we were to what we have the capacity to become. The suggestion is that the human species is a mere way station in a process leading to its self-transcendence. And, importantly, what gets transcended along the way is our corporeality and the limits imposed on us by our embodied condition. We are told, in effect, that we are spirits in the material world about to be released from the prison of creaturely existence.[7] The insinuation that our release from the Platonic cave of embodied existence is imminent is endlessly reinforced in popular culture and speaks to the deep cultural hold of a metaphysical or two-world worldview.

The fantasy that we humans are destined for release from our corporeal mantle explains the attention directed to "minds" in the literature: human minds, machine minds, the merging of minds, the ensuing meta-mind, and allusions about meeting the mind of God. In their mind-drunkenness, the clergy of the technology-will-save-us faith indirectly reveal what drives their apology for humanity's liberation through technology: namely, an aversion to the world's materiality. What grounds their unrestrained embrace of technology and technological innovation is a whole-hearted rejection of the worldly character of human existence. These technophiles are gnostics at heart. The material world is the locus of evil for them, and they cannot wait for technology to deliver humanity from a corrupt and corrupting nature. The irony here is that our coming to grips with reality is exactly what this vision of technological progress undermines. Nothing would be more unreal than living in a Kurzweilian "Singularity," or any comparable end-state where we shed our mortal coil and live on in a nonspace, nontime continuum like the gods.[8]

It might be thought that by fixating on a reading of technology as idolatrous as Kelly's that I have set up a straw man that misrepresents by overstating the otherworldly (or cosmic) character of the technological enterprise. I beg to differ. As radical as they may appear to be, the virtue of works of this type is that they forthrightly promote a vision of technology typically soft peddled by pundits, circulated by the media, and absorbed by the lay public. This vision, shorn of its extravagances, informs us that technology is an unstoppable force. It is bigger than human, if not downright cosmic in some vaguely mystical way.

Consider as evidence an advertising insert in a popular magazine on the subject of The Internet of Things (IoT). It is scarcely different in its messaging than scores of other promotional materials issued daily in popular publications on the next technological revolution. It reads, in part, "The truth of the world today is one of functionally limitless information. Sensors of all varieties are ubiquitous and creating raw data at a rate that is simply inconceivable. The wonder of the IoT comes from harnessing that data, processing it, and serving the relevant portions of it to people in a way that frankly seems like magic."[9] Technology is sold here in a way that underscores, in typical fashion, its supernatural or superhuman powers. The Internet of Things isn't just another step in the evolution of tool use. No, it is a "wonder" that expresses the realization of the "truth" of today's world. And in its limitlessness, this harnessed information will serve us in an appropriately impressive way—magically.

The fact that, like all advertising, most technology-related advertising is hype is beside the point. What is most important is what technology represents to us, because, in the final analysis, perception is reality. Hype works. Fantasies sell. The narrative that technology possesses a transformative power of superhuman reach is not an outsized caricature of technology's "real" power. This narrative constitutes the core of technology's ideological vision. Technology, we sense, may be our undoing as a species or it might save us from ourselves, but it is no trifling thing, no inconsequential force in our lives or in world history.

It cannot be rightly concluded from the analysis above that technology at bottom is about relieving the human estate. One cannot, for instance, account for current biotechnological efforts to manipulate the human genome in these terms. Something else is going on here. The human estate is not for us something from which we seek mere relief. To relieve, after all, is to lessen or alleviate, to make acceptable an existing condition of hardship. We have long passed the point of seeking through technological advance merely to ameliorate the negatives associated with the human condition. Rather, we see in technology a transformative and liberating power capable of radically remaking the world, of transcending the given.

A world where, as Kelly advocates, the technological ethos is fully loosed is uncomfortably close to our own. Hardline technophiles may want us to pick up the pace of technological advance, but their preferred path is our path, their end our end. It is a world built according to the vision that there are no limits to what can be humanly accomplished. The notion that there may exist insurmountable boundaries capable of thwarting the technological dynamo's forward momentum is anathema to the Kevin Kellys of the world. They embrace what they perceive is technology's totalistic thrust, its insatiable appetite to transform the planet and everything within it into a "standing-reserve," to borrow from Martin Heidegger.[10] There is no middle ground for them, no way to conceive technology as coexisting with views and practices grounded in an ethic at odds with that of rational reordering.

Those who approve of the indefinite and infinite extension of the technical order defend their claim on the grounds it leads to human betterment. The assumption is that no other option, or combination of options, is capable of delivering the desired result. The key linking technology with human improvement is "change." As Kelly asks, in defense of the technium, "How else are we going to change?"[11] Human advance for him is equated with changing for the better, and this change in turn is inconceivable outside the realm of the technical. The assumption that human betterment is necessarily fabricated is deeply rooted in our culture. It is self-evident to us that the future is something we make and that, if made well, our lives will be improved as a consequence.

The "good life" as envisioned by technology has a character befitting the technological ethos. Because technology is not about anything external to its own dynamic, the "good life" as understood in a technological age can be defined only as a kind of numinous superefficient state where things work really, really well. And is that not the picture of the world Kelly tries to convince his readers is ideal, a world where our perfect machines merge perfectly with each other and ourselves to form a perfect totality? The merging of atoms and bits is our business. Kelly's vision of our evolutionary endpoint as a species simply extends a dynamic already in place to its logical, if absurd, conclusion. What is progress for us if not our getting out of the way of the technological dynamo by ceding to the demand to create ever more functional versions of ourselves, our environment, and the interface between them? What is progress but a protracted act of self-transcendence through technological means, a self-imposed regime of self-perfecting that seeks to cleanse reality of its imperfections? What possibly could be wrong with the picture of reality that guides us toward the creation of such a future?

Well, a lot, truth be told. As previously argued, the technological vision is severely blinkered while feigning omniscience. Its proponents wonder why anyone would fret over a global monoculture that affords us limitless possibilities. If a single way of life promises us everything, why the fuss? From this angle, the so-called "tyranny of technology" is a myth propagated by those either fearful of change or willfully blind to technology's infinite potential. However, as with any narrative, considerations not deemed storyworthy are conveniently excluded. The time has come to consider what might be missing from the technological narrative. How can it be said that technology offers us an impoverished reading of reality when the evidence seems to suggest the opposite, and when so many of us align our betterment with the ideals that infuse the technological dream?

To proceed with the analysis requires first some stocktaking. The selfstated objective of this study is to think about technology. This thinking, to recall, aims to gain some critical distance from technology as a means of assessing its impact on our lived experience of the world. It is simple enough to account for technology's power to shape our lives when the accounting is done from within the parameters of technological thought. We noted that is the approach taken by virtually all popular literature on the subject today. As argued, these assessments inform us of the innumerable ways technological advance is making our lives more efficient, more convenient, more limitless—in a word, more technological. Whether this advance comes at too high a cost in other domains of interest to us is openly debated in some of the literature. Can or should the negatives associated with technological progress, such as the encroachment upon privacy rights, for example, be managed through legal or other means is an ongoing concern in certain quarters. But regardless of the verdict, debates of this kind are averse to concluding our "going forward" may result in irrevocable losses. It cuts across the modern grain to think in such terms. Because we conceptualize history as linear and progressive, the claim that the quest for perfect functionality may backfire in a serious way is discounted as the product of an unreasonably gloomy mind.

A dismissal of this sort is next to meaningless, however. It tells us more about the tenor of our own thinking than the potential for actual regress and only serves to limit needlessly the range of valid questioning. So, in the spirit of open-mindedness, let us give thought to the unthought. Let us go where our instinct as moderns tells us we ought not. Let us avert our gaze from what technology does or does not do for us, and concentrate for a moment on what technology is doing *to* us. What impact does technology, in its guise as an idea and tool, have on its creators? What happens to "the human" in the context of a technological order? The short answer is that it disappears. We end up adopting the mindset of a Kevin Kelly who, we have seen, abandons the notion that what it means to be human can be thought outside the realm of technology. Whereas once it may have been possible to do so, he would have us believe the powers of technology have grown to the point where no such distinction is now tenable. We are thoroughly technological beings: technology is inscribed on and in our minds and bodies, and in our living environments. We humans have been remade in the image of what we have made, which means we humans no longer can lay claim to being distinctly human.

There is considerable academic literature on the theme of posthumanism. [12] It is not my intention here to provide an overview of this material. What I want to do, instead, is address the subject in the context of my analysis of technology, as already laid out. Now, to argue that technological progress undermines a characteristic central to our humanity assumes a prior understanding of what is meant by the term "human." This assumption presents us with a considerable (but not impassable) problem. For it seems a quirk of being human that identifying perceived threats to our humanity comes easier to us than securing a definition of what it means to be human. History is littered with efforts at self-identification, after all. Even accepting we have a nature—a contestable claim in itself—our understanding of what constitutes this nature has undergone numerous revisionings. Reason, faith, and labor, to name a few, have been linked over the centuries to what has been deemed fundamental to being human. More recently, a scientistic understanding of the human has rendered more problematic still any effort to encapsulate a fixed trait or capacity with the human.

Paradoxically, then, an inquiry into posthumanism cannot start with an analysis of what it means to be human. The situation requires that we bracket for the time being the question concerning human nature and instead focus on the more readily discernible sociological context of the modern world, which has been identified here with the technological order. Whatever speculative insight might be forthcoming regarding our nature, the surest way to this end is to begin with an examination of "the human" in its current technological context.

If, as claimed, the human condition currently is unthinkable outside the context of technology, then our thinking about what it means to be human is deeply inflected by technology, the ramifications of which must be taken into account if we wish to gain greater insight into what it may mean to be human. Only by attaining an awareness of what it means to think technologically, and to think "the human" technologically, might we be able to think the human anew and in a way that does fuller justice to the experience of being human.

A substantial portion of the debate so far has been devoted to understanding the import of thinking about technology. Central to this analysis is the fate of the human. On the one hand, we have postulated the consequences for humanity of thinking technologically. Kevin Kelly, we have seen, takes the position that the technological good is aligned with the human good and that technological progress is facilitating the realization of our true nature. This account of technology has been identified with the majoritarian view, which sees in technology the promise of self-fulfillment.

Unsurprisingly, the mind-set associated with the technological worldview is ill disposed to seeing the course of future development in anything but a glowing light. Technology promises to continue to give us more of what we want in terms of possibilities and opportunities, a promise of undeniable appeal. Thinking about technology, on the other hand, has us question the premises underlying the happy identification of the technological good with our own. What is questioned is less technology's capacity to produce self-fulfilled lives than the identification of the human good, the good life, with the pursuit of a tensionless existence. Technology, on this account, is validated on the grounds that it satisfies desire. It progressively removes the impediments to living the kind of life we ostensibly all want to live, a life of fulfilled desire, a life like Jon's in the Facility. There is not a whiff of limits in Kelly's ode to technology nor any like-minded defense of the technological status quo. Technology facilitates our humanity by virtue of its unbounded potential to erase the tension between the world as it is and the world as we would like it to be.

Clearly, technology opens up avenues of possibility that previously were closed to us. It hardly needs stating, for example, that through technological advance innumerable perceived natural injustices (such as disease) have been overcome, which has allowed for the fuller development of individual potential and for which we can be thankful. But it remains problematic that our humanity is tied to our capacity to rid the world of its ingrained injustices. If it were, then we could confidently assert that every successive generation in the modern era is more fully human than previous generations, a risible claim if there ever was one. It may be that the happiness of persons is linked to the realizing of their potential, but what makes us distinctly human strikes me as having little to do with the happiness quotient. Although more will be said on this important theme later, I want to propose here that technology is dehumanizing to the extent it abstracts from our lived experience of the world. That is to say, technology has a way of distancing us from our own embodied, subjective point of view, and in a way as to have us lose sight of the ground of our being. As we merge with our technological creations and their systems of interconnection, we enter a free-floating realm marked by the erasure of the distinction between subject and object, or self and other, a distinction arguably connected to our self-understanding as human.

It is not entirely evident how the arc of technological development might not only not fulfill the human promise but work against its realization. To clear up matters, however, requires first stepping back and revisiting momentarily the question of what constitutes our humanity. As already observed, no one can harbor any illusions as to its answer. There is no definitive or uncontestable response to this question and there never will be one. No amount of reasoning will convince Kevin Kelly, for instance, to accept what is proposed here as an answer. But to offer an alternative to the view that hitches our humanity to the star of technological prowess, let us consider as a general claim the following: To be human is to engage the world in a way that brings out fully our latent capacity for engagement.

"Engagement" is the pivotal term in the above assertion. The etymology of the word reveals that to be engaged with something is to be committed to contest that thing. Engagement therefore supposes an antagonistic entanglement of sorts. The phrase "engaged in battle" sums up the sentiment well. You are engaged when you are "up" against something. When engaged, the object of engagement commands your attention because it refuses to conform to one's designs and advances. It follows, according to this reading, that an engaged life has to do with leading a tensional existence. Let us park this idea for a moment and establish a few basic empirical and experiential facts about our worldly condition as a means of understanding more fully what is meant by a tensional life. What better place to start than at the beginning—with our natality.

Without prior consent, all of us are born into a world whose existence we had no part in creating. We are thrown into this world with an abruptness that resembles an entry into alien territory. But this world is not alien in one important sense: it most certainly is made of the same stuff as we. In fact, it should strike us as incongruous that the substance of the world is not constituted in a way that makes the emergence of sentient life inevitable. Whereas, to take a contrary example, no amount of time spent recombining an infinite number of LEGO blocks will produce a human being, there is something about the nature of nature's building blocks that has allowed for the emergence of conscious existence. For this reason, arguments in support of "natural teleology"[13] appear well founded. Yet for all this the world we are born into does not comply (at least not fully) with our all-too-human demand for meaning. For those with a historical sense (or are otherwise sensitive toward the variability of worldviews), the world seems to withhold from us its sensibility. As stated earlier, the world conceals as it reveals, and there appears to be no way of doing an end run around this barrier to complete understanding. In this important way, then, the world seems "other" than human and a very alien place indeed.

So we are confronted with a conundrum. What is the true status of the real and our placement within it? Are we at home in the world despite its apparent indifference to our interests and concerns? Or are we strangers in a strange land despite intimations of alignment? Or, perhaps most unsettling of all, do we inhabit a world that is simultaneously close and distant? Absent an argument capable of settling the debate definitively either way, it seems reasonable to side with the last option. An experience-based case can be made that the world neither coheres as an intelligible whole of which we are an integral part nor signifies an utterly meaningless concatenation of random elements of which we are one. Rather, reality is both, and we humans are suspended in the tension between being at one with and utterly apart from the world or between absolute meaning and utter meaninglessness.

With this realization comes a kind of wisdom missing from the modern perspective and one that brings us closer to the ancient Greek outlook. It informs us our placement within the order of things facilitates our comprehending its general contours, and more. Yet, at the same time, this knowledge instructs us of the limits of our understanding. It makes apparent the fact that the condition of possibility for total or objective knowledge is not at hand. Being in and of the world precludes the kind of "immaculate perception"[14] (to borrow an expression from Friedrich Nietzsche) required of objective understanding. Being part of the whole prevents our gaining the kind of distance from reality needed to understand the totality of existence. This inbuilt limit to cognitive perception lends the perceived world an object-like character, a phenomenological density. It ensures the world that opens up to us remains opaque or less than fully transparent to human understanding. As noted previously, this limiting factor preserves the mystery of the real by revealing that reality always transcends any given understanding of it. To conclude, being human, from the perspective outlined here, requires in part being attentive to the perceptual ambiguity that attends our placement within the whole of reality. This attentiveness supplies us with our distinctly human spirit, a questioning or erotic spirit that as far as we know no other living creature possesses in such a heightened fashion.

Now back to the issue of dehumanization. I have suggested that thinking about technology leads one to entertain seriously the notion that technology is a dehumanizing force in our lives today. How so, given what I have just said regarding our humanity? The answer lies with the fact that a technological society functions in a way to undermine the object-like character of the world—its otherness.[15] Technology's power lies in its capacity to de-objectify, to rend null and void the thing-like character of the world. To understand better this claim, it is important to note that the word *object* is a composite of the Latin *ob*, meaning "against," and *iacere*, meaning "to throw." So an object refers to something that stands in the way of something else, or otherwise opposes it.

How is it, then, that natural order can be thought in terms of the object? In two basic ways, it turns out. One, as observed above, by the world's apparent indifference to the human desire for fixed meaning. Two, because the world as it is given to us—the natural order—stands in the way of realizing other kinds of basic human wants, namely, those related to our safety and security. Nature, by its nature, does not align with our more pragmatic concerns, at least not always. Bad things happen to humans: we get devoured by animals, overtaken by disease, dislocated by natural disasters, and so forth. And if through human ingenuity we manage to have nature submit to our designs, she lets it be known that this submission is largely gratuitous and therefore revocable. Bad things happen to wily humans, too. Spears snap and nuclear power plants melt down. Modern technology works to overcome nature's challenge to our claim to authority. After all, what does attaining power *over* the natural order mean if not the nullification of nature's capacity to stand against or resist the human will to mastery? The object or thing-like character of the world is what gets in the way of making the world *our* world. So it must be vanquished. The ultimate goal therefore is to eradicate pushback in the quest for mastery. But triumphs always come at a cost, a point we moderns usually are unwilling to concede. In this instance, the cost is the disappearing subject, for as technology works to eliminate the object, it also effectively undermines that which the object stands against—the subject. As figure needs ground, the subject is sustained in tension with the object. Working to eliminate one pole of the tensional field simultaneously works to undermine the other.

This digression, I hope, will help illuminate the principle at work in the arc of technological development. The claim here is that technology tends toward the slackening of the subject/object polarity. In its attack on the object, technology also affects the disappearance of the subject—the human subject. How this two-fronted attack plays itself out in the real world of technology is the question of the day. We can begin to answer it by reiterating that the technological vision takes as axiomatic that everything can be brought under the aegis of the controlling mind, including the power that does the controlling. So it is that we impose upon ourselves the same regimen foisted upon non-human nature. The puppet and puppeteer are one: we are both subjects and objects to ourselves. But since technology's proper end is de-objectification, the object in us requires extinguishing as well. Gaining control over nonhuman nature is inseparable from the task of gaining control over the nature in us. So it is not enough, for instance, to create a technology that helps overcome the challenges associated with human mobility. The invention of the modern motor vehicle or the airliner needs to be supplemented with the invention of the means to integrate navigators into navigation systems, itself a means to the greater end of maximizing the efficiency of the task at hand.

There are two consequences that flow from this sort of technological advance and from technological advance more generally. The first is the rather unremarkable fact that technological invention frees us from the limits imposed upon as strictly natural beings. In the instance referenced here, technological advance has enhanced considerably our natural capacity to traverse space. By extending our powers this enhancement might be seen as empowering the subject. The picture is complicated, however, with the second of the two consequences of technological development. Inexorably, the progressive enfoldment of human into machine leads to the disappearance of the former into the latter. The driver becomes the driven, the pilot the flown, the purchaser the purchased, as we are incorporated into performance systems designed to heighten the effectiveness of whatever function we seek to realize.

The reason we are squeezed out of the technological equation is simple enough: We humans are lousy functionaries, an issue that will be taken up again in a subsequent chapter. For now we can note that strictly human competences arguably are less robust and reliable than those of our creations. From a functional perspective, we humans are substandard. We are not dependably rational creatures, have limited amounts of energy, and possess traits not always conducive to coordinated action. From the standard of operational efficiency, it is unacceptable that the "subject" be permitted to jeopardize a system's proper functioning. So it is the object within us—whatever within us that accounts for our propensity to screw up—that compels our disappearance into the system's circuitry and our ceding to its authority as an operational entity. Of course, one of the ironies of this development is that the effort to save ourselves from our incompetencies only serves to aggravate them. Ours is the age of de-skilling, where increasing numbers of persons have matured in a social environment that discourages the acquiring of physical and intellectual competences that only a few generations ago were considered mandatory. Smart technologies really do breed inept humans.

As our lives become more technologically textured, the view that we are better served as bystanders to our own mediated interaction with the world takes firmer root. We experience the world less from the inside out, as subject-actors engaged with the object-world, and more from the outside in, as if we are nodes in a systems-centered web. The upshot, the novelist Robert Musil noted almost a century ago, is that it appears naive to assume the most important thing about experience is the experiencing of it.[16] The gist of this observation is revealed by returning briefly to what was said about technology and the erasure of the subject-object polarity. To speak of this duality is to underscore the reciprocal nature of our interaction with the world. All acting is necessarily also a being acted upon. The subjective experience of touching something, for example, is one with a recording of the experience of being touched by that thing. Without an object to oppose, there is no sensory experience called touching. The same reciprocal dynamic applies to all forms of human-world intercourse. So if, as argued, technology seeks the dissolution of the object-world, it undercuts the experience of the bumping up of subject and object. And if the object can be said to contain or limit the subject, to give it form as riverbanks do to the water that flows between them, then its dissolution results in the subject's dispersal or diffusion. That we see ourselves today as nodes or relay stations in a systems-centered universe is evidence of this dispersal. We are resource units, information packets who exist by virtue of our commutation with other units in a field of total exchange. Less and less, under this dispensation, are we inclined to see the world opening up before us, as subject-actors, in a way that demands our direct and active involvement. Less and less does the world appear to us as erotically charged, as something that commands our attention and engagement simply by virtue of its obdurate presence.

A brief examination of a generational pedagogical trend might help supply a better sense of the dynamic under consideration. So-called student-centered learning, or learner-centered education, has been offered as an alternative to the top-down educational model of old that has instructors inform learners about matters that they, the instructors, deem worthy of being known and reflected upon. The claim is that the educational experience is more effective—students learn better—when students take more active control of their own learning environment. It needs hardly be said that a true learning experience is necessarily dialogic. Pontificating never works and effective instructors always encourage independent thinking. But in response to the challenge of the alleged superiority of a student-centered approach to learning one can argue that, far from being the countercultural corrective to the status quo, the bias dovetails perfectly with the technological ethos. What, in the final analysis, could be more technological in spirit, more in keeping with the wisdom of the day, than the ideology of self-empowerment that subtends the student-centered educational model? What could be more in tune with the technological zeitgeist than to claim learning can best happen only when the learner is shielded from the limits of his or her understanding, by sidelining the role of the instructor? *Eros* has no place in an ego-stroking environment where actors are emboldened to "take charge." The "it's all about you" mentality discourages encountering the world in a way that elicits understanding as a response. As Socrates well understood, knowing you do not know what you think you know is a precondition of learning. Being apprised of the limits of one's understanding is the gateway to greater understanding. To the extent ignorance is learning's "object," it is a fundament of intellectual engagement and productive of understanding.

To generalize from the illustration above, we can say that what draws us into the world (and therefore has the power to make a claim over us) is whatever eludes our grasp or fails otherwise to do our bidding—the world as object. It follows that with the object's progressive erasure the subject disappears for want of something to make a claim over it. Like oysters in an oyster shell, our comfort zone is such that we cease to perceive ourselves as existing in opposition to the world.

There is a good deal of hyperbole in the claim that we exist in a non-oppositional state. But we are working toward this end, and that is the point. Understanding how technological progress redounds upon the agents of development is a primary aim of this study. Making explicit the largely unspoken narrative of technology is the goal. What is the nature of the unfolding story that has us aligned with the technological project? Arguably, it is the story of deobjectification. Progress for us is measured by the degree to which reality conforms to our designs for it.

Alternately, this progress can be described as the movement from a resistance-based experience of the world to a compliance-based experience. As noted, technological advance in large part is associated with progress in efficiency and efficiency is measured in terms of pushback to the powers of mastery. The less resistance an operational system experiences in realizing its function, the more efficient its functioning and the more refined its performance. So resistance is bad and compliance good. The more we have of the former, and the less the latter, the better according to the lights of technology. The conclusion to be drawn is that the drive toward increased efficiency has a profound effect not only on the evolution of technology but also on our evolving relationship with our artifacts. Take the progression of automotive technology as an illustration. As automotive technology evolves so too does the automotive experience, and in a way that reflects the nature of its development. The difference in experience this evolution effects is the difference between "commanding" a vehicle to perform as one intends and "participating" in an operation whose end is the co-management of a function. The latter experience is technical. Driving today is fast becoming a matter of monitoring the monitors and overseeing the efficiency of the operating system nominally commanded. The skill-related demands placed upon its human operators are continually being minimized as the evolution of driver-assistance technologies relieves us of more and more of the work of driving. The self-driving vehicle is the apotheosis of the technicizing of the motoring experience. Emblematic of a more general trend, progress in automotive technology is making humans redundant. We are building for ourselves a world where human competences do not matter.

It is instructive to note that, in its original sense, the word "drive" means to push from behind. The act of pushing, in turn, implies the existence of a counterforce, a force of resistance—an object-force, as it were. We know, for instance, nails do not drive themselves into boards for the simple reason that it takes force to do so, or human agency. And it is precisely the felt resistance to the act of driving a nail that, if carefully attended to, enables the cultivation of the skill of nail driving. The same applies to any other learned competence. It is the object-like character of the world that allows for the development of effective human agency. Yet while resistance may be considered a good when viewed from the angle of human competency, it is a good whose value is trumped by the good understood technologically. Judging from the trajectory of technological advance, it is machine competence we value more than its human counterpart.

So, to return to our original example, what becomes of the driver and the driving experience when driving becomes driver-proof, when a vehicle ceases to exist as an "object" of manipulation? They disappear. They become superfluous. The redundancy of human competence is happening across the board. We are becoming bit players in sweep of things. Consider another example we are equally familiar with: passenger air travel. How is it possible to sit in the latest iteration of an airliner as it rotates off the runway (on a computer's command) and not be struck by the uncanny feeling that the experience of "flying" has effectively disappeared? Airline pilots do not fly aircrafts: They manage the systems that fly aircraft, along with the myriad of other personnel in airport lobbies, control towers, and maintenance hangars. And we, the paying customers, are not flown. We do not want to be flown. Rather, we want to be expedited from place to place and in a way that allows us to do much the same thing we do when stationary—text, watch a film, write a report.

The general conclusion to be highlighted here is that the arc of technological development favors the subject-less experience over its antithesis. We are being conditioned to be cool with being cool. If the logic of technological mastery sees us progressively excised from the work required for a full-bodied and able-minded engagement with the world, then so be it. The benefits outweigh the costs we tell ourselves. We live safer, more productive lives by assenting to our immersion in, and our reliance upon, the countless systems that navigate us through our daily activities.

It is worth reconsidering George Saunders's "Jon" in light of the discussion above. As noted, this narrative satirizes the worldview upon which the technological project rests. The functionaries that populate the story do not desire or want for anything; they perform or act out. They do not yearn; they acquire. They do not suffer; they realize. And if these ends cannot be satisfied, if the gap between desire and its satisfaction cannot be closed by satisfying desires—if, in short, reality intrudes, as it occasionally does—the same end can be achieved the other way around by readjusting desires artificially to make them satisfiable by existing means. Either way, Saunders informs us, by consumerist or pharmaceutical means, we are reconstructing the world in a fashion that suppresses the subject by satisfying desire. The self disappears when it is touched by nothing, when it ceases to meet up against anything that forces the realization that because the world is not infinitely manipulable, the world forever will remain "other" than the self.

In "Jon," and for us, the manufacturing of self-satisfied beings is considered praiseworthy because desire is the enemy of social stability, and social stability stands among our most valued collective ends. The presence of desire indicates dissatisfaction with the given, a theme as dear to George Orwell as George Saunders.[17] For desiring or erotic beings, the world as it is is not the world as it might be. The real and the ideal are differentiated, and remain differentiated. So desiring beings are not "at one" with their environment: they remain existentially unsettled. I have suggested that this unsettledness is a vital component of what makes us human. Not being at home in the world is pivotal. As Saunders reveals so effectively, that the world is constituted in a way that frustrates our desire for self-satisfaction is not the problem. To the contrary, the problem today lies with the effort to erase the tension between what we want and what is on offer.

In conclusion, it bears repeating that appreciating the object-like character of the world orients us to the order of things in a way that opens us to the world beyond our heads. Understanding that the world is not our world, but *the* world—in many important ways an *alien* world—forces upon us a recognition of the thereness of reality, or the "givenness of things," in Marilynne Robinson's favored language.[18] The world stands before us, and we before it, in a way whose reality is obscured when increasingly our contact with the world is technologically mediated. Like artists who train themselves to register what they see in a way that precludes, as much as possible, what they have been conditioned to see, today we must work at cultivating an appreciation of the sheer facticity of what surrounds us.

At our peril we take for granted that which resists or opposes us. When we rework the world in a way that dissolves its object-like character, we are inclined to live in our own heads, to live introverted lives. The outward gaze is redirected inward when the world loses its capacity to charm. We lose our perceptual faith in the reality of lived experience. We learn to distrust the appearing world as it is revealed to us directly through the senses and our self-reflective capacity to understand the world and our relationship to it. We become less naive, less Socratic, in the belief that through unaided reason we can make sense of the world, and more technocratic in our reliance on systems of knowledge that aim to conform reality to our sense of it.

What better symbol of modern introversion (and the disappearance of the subject) than the downward glance of the ubiquitous texter? It is time the iconography of Raphael's *The School of Athens* was updated. The transcendent and worldly gesturings of Plato and Aristotle need to be supplemented by that of an archetypal texter, posing midstride, shoegaze-style. This complement would exemplify our allegiance today neither to a world beyond the natural order, nor to the natural order itself, but to a surrogate reality that is the technological order.

Missing from the technological vision is the centrality of the lived experience of everyday reality to our self-understanding and our broader understanding of the nature of reality. Such experiences are devalued because they are seen as incapable of supplying insight into the nature of things. If, arguably, the disappearing subject marks our entry into a posthuman age, it is a result of the radical devaluing of the insiders' view of reality and our own displacement from lived experience. Too much has been gained from adopting the Archimedean point, the exteriorizing perspective, to see what has been lost by the abandonment of the subjective experience of the world. We are so busy pushing things around we fail to appreciate to what extent we remain conditioned by forces beyond our control. That the object-like character of reality cannot be wished away is a given. Yet we moderns attempt to do just that and are living the consequences. As our technological hold on reality extends and tightens, our ability to envision an alternative vision to the technological decays, only further to secure our technological fate.

Kevin Kelly and his ilk are not to be dismissed for their portrayal of the developmental pattern of technology's evolution. Technology is indeed an abstracting force that buffers humanity's encounter with an unforgiving nature. Rather, the concern is with the normalization of the vision that tells us we are well served by linking our fate with the technological dynamo. What requires explaining is how we have arrived at a point where it is assumed *human* betterment is facilitated by our disappearing into the circuitry of technological systems. How has the problem with reality been couched in terms sufficiently radical to have the likes of Kevin Kelly be regarded as forward-thinking humanists? To answer this question necessitates addressing the problem with "the problem with reality." It requires examining the false ontology that underlies the modern vision, a misunderstanding of the nature of reality responsible for an unthinking embrace of the technological project.

It goes without saying that solutions to problems reflect the way problems are framed and perceived. If technology is a solution to the problem with reality, it represents a solution fitted to the perceived nature of the problem. In an attempt to appreciate better the relational character of problem solving, consider for a moment the "state of nature" debate that raged in the West during the seventeenth century. Thomas Hobbes, a prominent participant in the intellectual tussle, is known for his notoriously dark depiction of what humans are "by nature," or outside the institutional framework of the state. By trying to convince his readers of humanity's natural antisociality, he defended the need for an artificial corrective of equal radicalness, namely, a sovereign power with near total command over its subjects. Another social contract theorist and fellow countryman, John Locke, rebutted Hobbes's portrayal of "natural man" with a view that took the state of nature to be less hellish than troublesome. Accordingly, his solution to the problem of the natural human condition was significantly more benign than Hobbes's: representative government sufficed for Locke. In both instances, the solution to the problem at hand matched the nature of its depiction.

As will become more apparent shortly, the technological worldview supports a vision of reality that, unsurprisingly, sees the implementation of ever more effective systems of control as its solution. The solution makes eminent sense given the existing description of the problem it intends to fix, but decidedly less sense in the context of an alternative description of the nature of the real. It is time, then, to supply an alternative vision of reality—a description of the real that purports to better fit the nature of reality—as a way of further distancing ourselves from the ideological hold of technology.

It has been suggested that by putting a premium on values such as control and efficiency a developmental agenda has emerged that suppresses a key feature of our humanness—our erotic nature. It has been argued that this drift manifests itself in our ceding to technology and its governing ethic a determinative role in the unfolding of societal development. This is not simply a theoretical claim. We *live* decentered lives. We experience daily the consequences of our collective decision to heed the call to operational efficiency. We need to look no further than our work lives to understand how profoundly we have adapted and readapted to the vagaries of a globalized economy. Workforce mobility is a paramount virtue in an economy that continually shape shifts in response to the evolving demands of a networked commercial world. You do not have to be a socialist to appreciate Karl Marx's observation that the constant revolutionizing of production in the modern era has a corrosive effect on all "fixed, fast-frozen relations."[19] The system is controlling, the performance principle in command, and we go along for the ride making whatever concessions are necessary to ensure a realigning of the human good with the good of the system.

Required now is a further exploration of the linkage between technology and the disappearance of the subject, on the one hand, and the picture of reality that sustains technology, on the other. If, as argued, the world we make reflects the world as we understand it, then securing a fuller understanding of our efforts at reconstruction demands a closer examination of technological world-picture. To do so requires a brief return to a subject touched upon earlier, namely, the phenomenology of perception.

It already has been established that perception is perspectival. This means, to repeat, that the perceiver is never in a position where the perceived reveals itself fully. To perceive anything in its totality presupposes the adoption of an all-seeing vantage point. This vantage point in turn presumes the seer is at a fundamental remove from the seen, or alienated from the realm of the visible: God-like, in short. That no such vantage point is attainable with natural perception means the world is incapable of revealing itself fully to our gaze. There are at least two reasons why. First, as stated, we always perceive reality from a determinate position *within* the world we envision, and therefore the world we see is limited by the horizon of our situated vision. Second, human vision is not x-ray vision. We perceive surfaces and to the extent we do the world reveals itself to us as a concatenation of profiles. So as embodied beings we are never in a position to see everything at once, and what we do perceive we perceive only suggestively. Yet despite the inherent partiality of perception—its incompleteness—the perceived world takes on the character of fullness, as Maurice Merleau-Ponty has taken pains to point out.[20] The visual profile appears filled out: it appears "real" in that sense of the word we all intuitively grasp but have difficulty articulating.

How the perceived realness of the world is generated remains an open question and beyond the scope of this inquiry. It is sufficient to note, however, that developmental psychologists have explored the issue in depth and have concluded that, in general terms, it is in response to our active bodily engagement with the world that we gain a sense of its existential presence.[21] By means of visual exploration, for example, a child comes to understand that a cube has more sides than can be observed from any single angle. By extension, we learn that the perceived world is always "more" than what is revealed to us at any instant. The upshot is that reality, for embodied observers, *transcends* its presence as an appearing thing. We sense or intuit this much without ever having to articulate it as a proposition about the nature of perception.

Again, we can ask what follows from the realization that the appearing world is always more than what it reveals itself to be at any given moment. Is there some kind of insight into the nature of the world that can be drawn from a phenomenological analysis of perception? Yes there is, and it has to do with the effect—readily accessible by us—of the complementarity between the seen and the hidden, between the visible and the invisible, that marks natural perception. This effect manifests itself as the seduction of the real. What draws the perceiver into the world, draws the eye and its gaze to explore visually what lies before it, is the interplay between the revealed and the withheld, or between presence and absence. In other words, while the perceived world is never complete—never total under the human gaze—it nonetheless evokes completeness and it is the evocation of totality that draws us into the world and seduces us. We can do no better than return to Saunders's description of Jon's first encounter with the real world, the world outside the Facility, to experience vicariously the seductive power of the real. Less really is more.

Arguably, the problem with the "less is more" ethic is that it holds only to the extent a person is attuned to the subtler dimensions of visual perception. Most of us are not phenomenologists. Nevertheless, most persons are attuned to this extra dimension even though they likely have not conceptualized the experience. It resides in the ineffableness of the appearing world, the sense that the perceived world exceeds its presentation or otherwise points to something beyond what can be absorbed by the five senses. This experiential surfeit is what captures our imagination and prompts us, for instance, to seize it visually by technological means, in the hope (always dashed) of securing the experience of perceptual transcendence.

It is easy to overlook this insight and to assume the reverse, namely, that more is more, especially in the age of the screen where images are ubiquitous and hyperreal. But therein lies the problem with the technological world-picture. The problem with technology as a worldview is its betrayal of the phenomenological insight that perception is perspectival and therefore incomplete. The technological picture of the world is objectless, unlike embodied perception. That is to say, the technological vision assumes nothing in principle stands in the way of complete or total vision. The technological vision is omnivorous: It admits of no hiding places. It assumes a God's-eye view of the world, that the hidden is simply the not-yet-visible. This pretense to the total gaze replaces the searching or erotic look, a perceptual switch that facilitates the marginalizing of lived experience and the wisdom derived from recognizing the fragility of all efforts to make coherent and complete sense of the world.

To highlight the contrast between these two modes of perception, consider for a moment that for most of our species' history weather watching was a distinctly embodied or first-person experience. Our forebears looked out into the world to discern signs of impending shifts in weather. As an embodied venture, weather watching entailed a multisensorial reading of the world, supported by generations of observational lore. Now we look down at images taken from space and read weather station data to gain a picture of the same. Efficacy aside, the gap in the phenomenological richness of the experience of weather information gathering between the prescientific and scientific eras is impossible to overstate. With the former, the skills required for the task at hand were a refinement of the observational capacities shared by all human beings, and were exercised in situ. With the latter, in contrast, weather reading is a mediated affair that calls on skills pertaining to the interpretation of weather data, of which human judgment plays an increasingly insignificant role.

We are all meteorologists today to the extent we adhere to the assumption that the more complete our picture of reality—the more total our vision of the real—the better the fit between us and it. The object is eradicable we believe and we have a duty to erase that which obscures our vision. It is all figure and no ground for us, all light and no shadow. Kazimir Malevich's 1918 painting, "White on White," captures perfectly the spirit of the modern age—the age of technology—by evoking the imbalance at the heart of its governing vision.[22]

It is now time to reexamine the technological ethos within the context of what has just been said. The technological will to power seeks an ever-expanding understanding of the world as a precondition for extending control over it. The feasibility of the enterprise rests on a presupposition regarding the "truth" of the world, namely, its intellectual transparency and amenability to control. As moderns, we assume the world will submit to our interventions. The technological venture is premised on this key assumption and would make no sense in its absence. But what if we got things wrong in this regard? What if the technological worldview, with its faith in the transparency of the real, misreads fundamentally the reality of reality? And if it is misguided in this way, what are the implications of this oversight for the real world of technology? All are important questions, and the subject in this chapter of our continued investigation into the problem with technology.

NOTES

1. After pondering the argument that God and luck control our fate, Machiavelli concludes in the penultimate chapter of *The Prince*, "Nevertheless, since our free will must not be eliminated, I think it may be true that fortune determines one half of our actions, but that, even so, she leaves us to control the other half, or thereabouts."

2. This assertion is drawn from Henri de Saint-Simon's "Letters from an Inhabitant of Geneva to His Contemporaries," in *The Political Thought of Saint-Simon* (Oxford: Oxford University Press, 1976.) "Letters" is available online at: https://www.marxists.org/reference/subject/philosophy/works/fr/st-simon.htm.

3. For a fuller account of Lawrence Krauss's position, see *The Atlantic* interview, "Has Physics Made Philosophy and Religion Obsolete?," April 23, 2012.

4. The expression "specialists without spirit" is Max Weber's and is found in his *The Protestant Ethic and the Spirit of Capitalism*, tr. Talcott Parsons (New York: Charles Scribner's Sons, 1958), 182.

5. My commentary in this and other chapters on Kevin Kelly's vision of technology is gleaned from his *What Technology Wants* (New York: Penguin, 2011).

6. Kelly, *What Technology* Wants, 355.

7. For a fuller discussion of the dematerializing effects of technological advance, see my *Spirits in the Material World: The Challenge of Technology* (Lanham, MD: Lexington Books, 2009).

8. The notion that the cosmos is evolving into a state of super-complexity has legs. In *The Phenomenon of Man* (1955), French mystic Pierre Teilhard de Chardin coined the expression "Omega Point" to capture what he claimed was the endpoint of cosmic evolution. In only a slightly less religious context stands Ray Kurzweil's "technological singularity," which he speaks to at length in his work, *The Singularity is Near: When Humans Transcend Biology* (New York: Viking, 2006). A tech version of the Omega Point, his Singularity signals the coming of a form of superintelligence. Kevin Kelly, similarly, adheres to the belief that humanity is destined to self-transcendence.

9. *Macleans*, April 11 (2016), 46.

10. In Heidegger's parlance, the truth today is "enframed" (*Gestell*), or given form, as "standing-reserve." Things are framed for us such that what we see as true and real takes the form of manipulable matter. See Martin Heidegger's "The Question Concerning Technology," in *The Question Concerning Technology and Other Essays*, tr. William Lovitt (New York: Harper Colophon Books, 1977).

11. Kelly, *What Technology Wants*, 348.

12. Classic texts on posthumanism include Donna Haraway's *A Cyborg Manifesto*, N. Katherine Hayles's *How We Became Posthuman*, and Francis Fukuyama's *Our Posthuman Future*. The list of more recent publications on the theme is long, but among the latest are: Neil Badmington's *Alien Chic*, Rosi Braidotti's *The Posthuman*, Steven T. Brown's *Tokyo Cyberpunk*, Stegan Herbrechter's *Posthumanism: A Critical Analysis*, Jennifer Koosed's *The Bible and Posthumanism*, Pramod K. Nayar's *Posthumanism*, and Cary Wolfe's *What is Posthumanism*? As a point of clarification, my interest in posthumanism (versus "transhumanism") is rooted in the fact that posthumanism speaks to what I take to be the logical endpoint of the transfigurative power of technology to transcend the human. Transhumanism, in contrast, addresses the transitional stage in the movement from the human to the posthuman.

13. Thomas Nagel is one prominent proponent of natural teleology: As he puts it, "I am drawn to a fourth alternative, natural teleology, or teleological bias, as an account of the existence of the biological possibilities on which natural selection can operate. I believe that

teleology is a naturalistic alternative that is distinct from all three of the other candidate explanations: chance, creationism, and directionless physical law." See Nagel's *Mind and Cosmos* (Oxford: Oxford University Press, 2012), 91.

14. Friedrich Nietzsche uses the expression "immaculate perception" to mock the notion that perception is clean or untainted by the object of perception. In essence, Nietzsche here advocates for an embodied understanding of perception. See "Of Immaculate Perception," in Nietzsche's *Thus Spoke Zarathrustra*.

15. My analysis of technology's role in undermining the object-like character of the world is inspired by Jean Baudrillard, a theme to which we will return in subsequent chapters.

16. The full passage, from chapter 39, of *The Man Without Qualities*, from which this notion is extracted, reads, "A world of qualities without a man has arisen, of experiences without the person who experiences them, and it almost looks as though ideally private experience is a thing of the past, and that the friendly burden of personal responsibility is to dissolve into a system of formulas of possible meanings. Probably the dissolution of the anthropocentric point of view, which for such a long time considered man to be at the center of the universe but which has been fading away for centuries, has finally arrived at the "I" itself, for the belief that the most important thing about experience is the experiencing, or of action the doing, is beginning to strike most people as naive."

17. The parallels between George Orwell's *Nineteen Eighty-Four* and Saunders's "Jon," I think, are relatively obvious. Chief among them is that both unfold as love stories in totalizing environments.

18. See Marilynne Robinson's *The Givenness of Things: Essays* (New York: Farrar, Straus and Giroux, 2015).

19. Karl Marx, *The Communist Manifesto*, ed. Frederic L. Bender (New York: W. W. Norton, 1988), 58.

20. The importance of Maurice Merleau-Ponty's phenomenoloy of perception to our ongoing discussion of technology will be addressed in further detail in chapter 7.

21. A major exponent of the connection between learning and worldly engagement is Jean Piaget. This association is especially apparent during the early sensorimotor stage of child development.

22. The title of Kazimir Malevich's "White on White" points to the artwork's minimalist signature. Inspired by the Russian Revolution, Malevich intended the painting to represent the coming of an age of spiritual freedom. With its emphasis on purity and spirituality (as opposed to gross materiality), "White on White" fits well with technology's ambition to transcend the world's corporeality—the locus of the object. Malevich's masterpiece is therefore objectless, both literally and figuratively.

Chapter Four

Evil and the Empire of Good

It is time to expand our analysis in a way that brings it into closer contact with an especially perspicacious reading of technology. To be consistent with the claim regarding the perspectival nature of perception, it cannot be asserted here that this alternative reading is a "true" reading. But it can and will be argued that it constitutes a vision more adequate to the reality it seeks to describe than the technological world-picture, and one against which we can assess the technological vision. The reading in question belongs to the French theorist Jean Baudrillard. [1]

Baudrillard reinvigorates the spirit of wonder that enlivens and sustains thinking, understood in its fullest sense. His animus toward technology is directed less at its material effects than its impact on the intellectual imagination. To the extent Baudrillard thinks about technology in order to save thinking, his intentions are in keeping with those of this study.

Despite the powers technology unleashes and the array of possibilities it opens up, Baudrillard understands well the limits of the technological mindset and how these limits have skewed our understanding of what lies within the realm of the thinkable. We moderns act as if all of history has been a prelude to the realization of what constitutes true thinking, which for us is synonymous with instrumental or technoscientific reasoning. The question of meaning has been settled, too. We have bought into the narrative that humans have been placed on Earth to prevail over it and that getting on with the job of domination is our life's mission. As observed, from within this cultural context thinking is identified with instrumental or means-ends reasoning. There is no room here for thinking "outside the box" if we mean by this idiom something other than what is meant today, namely, conjuring new solutions to technical problems.

Baudrillard shows us what it means to think beyond the governing strictures of the day. And if some critics downplay the seriousness of his intentions, it is more a result of his exposing the stultifying character of much of what passes for legitimate thinking in our age and less because Baudrillard's analysis of modernity lacks merit. Central to this analysis is his reading of what constitutes reality. Parallel to what has been argued here, this reading underscores the ineffable character of the real, or its object-like essence. It draws attention to the fact that reality is profoundly misinterpreted when we fail to recognize it as something interpreted.

It was previously stated that we fall into a trap when we assume the world is, in principle, intellectually transparent. That the world exists does not imply access to its univocal truth. We moderns err in thinking it does and for Baudrillard this is *our* problem, not reality's. Like literalists who read the Bible as revealing God's very word, we moderns overlook the symbolic character of the "text" called the appearing world. We forget the technological worldview, like any other, is a reading of reality, not the real itself. We forget that seeing the world as a fathomable and exploitable resource is how reality represents itself to us, not how it "really" is.

If we keep in mind the symbolic character of the appearing world, its status as a representation of the real, then only one conclusion can be drawn regarding its nature: it doesn't have one. Reality is an illusion, Baudrillard submits.[2] Clearly, it needs be said, Baudrillard has a flair for the dramatic. He is a Gallic provocateur. But he makes here an important philosophical observation, one that reinforces what already has been said here regarding worldviews. It amounts to underscoring the distinction between map and territory. The reading of reality (the map), he admonishes, is not to be confused with the real itself (territory). There is a gap between the two the existence of which prevents all of us from being Buddhists or atheists or Scientologists. So for Baudrillard the noncoherence of interpretation and text simply is. There is no wishing it away. As he puts it, "The definition of the world or of the universe is that there's none that could be expressed in its totality. There's no mirror in which the reflection of the world could be caught."[3] It is for this reason that Baudrillard concludes that belief in reality is one of "the elementary forms of religious life," and reflects "a weakness of the understanding, a weakness of common sense."[4] It follows that the technological worldview is the spawn of a fundamentalist or religious mindset generative of various forms of fantastical thinking. We moderns, Baudrillard maintains, are especially fond of the reality myth. We own this illusion and organize our lives around it. It is an article of faith for us that secures our way of life.

Baudrillard thinks we would be better off if the illusion of the real were accepted. Acknowledging this illusion means recognizing what has been called the object-like character of the world. It amounts to recognizing not only that the nature of reality remains a mystery but that it will forever remain so. It takes work to ignore the illusion of the real, in Baudrillard's estimation. It is all too evident, he suggests, that the world we are born into amounts to a crime scene whose perpetrator has pulled off what all perps aim for but none have realized—a perfect crime.[5] For Baudrillard, the irreducible otherness of reality precludes its leaving traces of its origin or purpose, or at least traces of the sort that might lead to solving the crime. What is behind it all? How can this crazy open-ended world of ours be? Reality is the coldest cold case of them all. So cold, in fact, that Baudrillard says we have no choice but to cease efforts to crack the case.

Baudrillard contends "all great cultures" have come to terms with this core feature of reality.[6] These cultures understood the futility of trying to convert illusion into Truth. Because these cultures were mindful of the difference between a worldview and the world itself, their thinking remained analogical or symbolic. *Pace* Plato, they did not strictly oppose myth to rational discourse, for instance, because they understood that the truth (such as it exists) can be grasped only darkly, through its images. This insight escapes us moderns. We like our truth neat and dismiss more nuanced expressions of the truth as no truth at all. Reality for us is a revealing and only a revealing.

Our modern-day obsession with "reality" indicates our collective incapacity to relate to reality on its own terms. It explains the earnestness of our efforts to get to the bottom of things, to explain how the world *really* operates, as a means of making it work better. For us, scientific understanding represents not just a mode of perception but insight into the true nature of things. The evangelical aura that attends the pursuit of technological power flows from this largely unquestioned conviction. As observed earlier, in this regard the technological mindset is indistinguishable from the religious.

If reality can be liked to an illusion, for the reasons just stated, then efforts to dis-illusion the world serve to remove us further from the real. Umberto Eco expresses this sentiment well in *Foucault's Pendulum* when Casaubon, the novel's protagonist, declares, "I have come to believe that the whole world is an enigma, a harmless enigma that is made terrible by our own mad attempt to interpret it as though it had an underlying truth."[7] For Baudrillard, the greatest crime pertains not to reality's illusory character, but to our efforts to dis-illusion the illusion, or to make reality real. We assume that by informationalizing, virtualizing, and reordering reality to accord with our notions of perfect functionality we are "getting real" as has no civilization before us. Kevin Kelly's account of technological progress perfectly captures this sensibility. The universe is waiting to meet us, he opines, and technology is the means of our mystical communion with reality.

The understanding that reality is illusory does not amount in itself to a worldview. If we were to stop here and agree that reality is an illusion we would be left in the position of a nihilist for whom there is no point in making any substantive evaluation about the nature of the real. The truth claim supporting the technological worldview might be shown to be inadmissible on the grounds all such claims are philosophically illegitimate, but nothing more could be said critically about this worldview and the social order that issues from it. Baudrillard, however, is a skeptic, not a nihilist. His disbelief in reality is epistemologically grounded. That is to say, he remains skeptical of the veracity of any truth claim on the grounds that the means of such discernment are not at hand. But this is not to say that some semblance of true understanding cannot be extracted from the given world. Reality for Baudrillard may be "a dog" in that it "bows to any conceptual violence."[8] It certainly is the case that for him reality's unreality issues from the world's overwillingness to submit to imputations of meaning. But it is important that reality's perverse indifference to conceptualization manifests itself through its amenability to meaning, not its withholding of meaning. For reality's indifference to meaning for Baudrillard does not preclude conceptualizing reality in a way that minimizes the violence perpetrated against it. Something approximating true meaning therefore is never out of bounds.

Baudrillard's skepticism has the positive effect of helping elaborate a typology of worldviews based on the distinction between those that respect the reality's illusoriness and those that do not. The technological worldview clearly falls under the latter grouping. Its claim to truth is proof. In contrast, a worldview that falls on the other side of the divide makes no claim that its vision of reality provides an exhaustive account of the real. It remains open to the mystery of the real. What might such a worldview look like? Baudrillard offers us a glimpse.

It is fitting that our unthinking faith in technology to secure a better world is based on an essentially religious perception of reality. That perception may be called Christian in its assuming the essential goodness of creation. This goodness is rooted in the belief that the world coheres at some basic level, or holds together as an ordered whole. It is important to stress that "at some basic level" the world hangs together because the world of everyday experience belies this goodness. So any assertion touting the *essential* goodness of the world must include a proviso explaining why the world does not appear to be good. Enter evil. Evil, within this religious context, can be seen as the counterforce that compromises good order. And, importantly, belief in the essential goodness of creation bolsters the view that whatever works against this order is eliminable. Baudrillard has a lot to say about evil, a topic of discussion that will occupy a good portion of this chapter's focus. For now, though, it is important to note that a central characteristic of the Christian narrative is that the presence of evil or disorder in the world is a sign of its corruption, and that the world's fallen state is redeemable in principle. The teaching tells us a bad world can be made right.

The teaching that informs us a broken world is repairable is shared by the technological belief system. The notion the world can be rid of that which corrupts it is a critical element in the technological worldview. Again, we see with Kevin Kelly a classic illustration of a mind-set that sees the march of technological progress leading inexorably to a perfect world where perfect beings do the kinds of things perfect beings do. Analogous to Karl Marx's depiction of a future classless society, the technological vision dangles in front of us a picture of a world devoid of everything that might interfere with the realization of our wants and desires. It is a world where limitless beings do limitless things—forever. It is a world where super-civilizations populate entire galaxies, making sweet techno-whoopee. It is a world purged of the object, a world redeemed and made whole.

In any form they may take, redemption narratives make sense only on the assumption they rest on a true account of the real. Baudrillard has the good sense to reject the technological version of the redemption yarn. He speaks with the conviction of someone who has seen the fatal flaw in the narrative. But Baudrillard says more than this. It is flawed for him not simply because it mistakes its picture of reality for reality itself, but because the content of the vision is wanting as well. The technological worldview misrepresents reality and Baudrillard shows us why by balancing the reigning world-picture against an alternative he suggests better accounts for the world.

The question now arises: On what grounds can Baudrillard advance an alternative vision of the real? On what grounds can anyone? The answer lies in the conviction that our subjective experience of the world can tell us something of importance about the world. In other words, we have the equipment, simply as sentient beings, to discern the contours of our existence. In this sense, Baudrillard's position resembles that of the ancient Greeks for whom unaided reason, well employed, has the capacity to reveal much about the human condition. His discernment leads him to conclude that reality is inherently self-divided and in a way that is unrepairable: the human condition, and the cosmic condition it mirrors, cannot for him be redeemed by any human or nonhuman natural force. Rather, the world is fated to exist in the tension between the forces of order and disorder. An analog of the subject-object duality spoken of earlier, Baudrillard contends that "evil" is an essential feature of the real, along with the good, and that the world is an arena within which the forces of order and disorder contend in a never-ending battle. He goes so far as to add that of the two forces, evil is the superior: "Evil rules the world; Good is an exception."[9] From the modern perspective, this means the problem technology intends to fix cannot be resolved. The object is no mere stone on the path, to be kicked aside as we make headway on our technospiritual journey. To the contrary, the object for Baudrillard is the Object, an ontological and hence a permanent feature of reality that asserts its power regardless of our efforts to wish it away or surmount it.[10]

It bears repeating that an alternative to the false ontology of the technological worldview cannot be a true ontology in any straightforward sense of
the word "true." Its status as true is a relative one in that it better speaks to
the world of common experience than the technological world-picture. This
appeal clearly is not testable as are scientific knowledge claims. If it per-
suades, it does so by appealing to our judgment about the nature of things as
evidenced by our reflections on everyday experience.

Baudrillard's ruminations lead him to conclude that the world does not
operate the way we moderns think it does, or the way we act as if it does.
Every time we get swept up in the hype surrounding the latest gadget's
promise to revolutionize the world, the belief is reinforced that technology
holds a saving power capable of delivering us from the "evils" of the world
that thwart the realizing of our dreams of perfection. Matters of truth aside,
we must concede first that the picture of reality presupposed by the promise
of perfection is contestable. A first step has been taken in showing what
might be deemed a crack in the technological world-picture. Whether this
crack is picture shattering depends on the extent to which the technological
vision actually misreads the nature of what it seeks mastery over. Which
picture better fits the reality of our experience, technology's or the alternative
alluded to here? The answer to this question is less important than the asking.
Thinking about technology provokes this line of questioning. It challenges
the viability of the premise that underpins the technological project. To this
end, it is time to revisit the real world of technological development and
examine more closely what our actions tell us about the world we think we
understand.

Murphy had it right: anything that can go wrong will go wrong. The
universe appears haunted by a perverse daemon intent on undoing what has
been done. At best it seems like a "two steps forward, one step back" kind of
world, and even then not for long. How we relate to this apparent "natural
law" is illustrative of how we view the cosmos. Our assessment of what may
be termed the "tragic" element of worldly existence reflects a cultural bias
toward what is assumed the "proper" course of things.

The ancient Greeks were obsessed by tragedy and made their amends
with it as best they could by accepting the contradictoriness of human existence. Their existential predilections precluded hopes of attaining what all
humans strive for, a world that complies fully with one's desires. As E. R.
Dodds says of the tragic playwright Sophocles, he expressed the "overwhelming sense of human helplessness in the face of the divine mystery, and
of the *ate* [folly, ruin] that waits on all human achievement."[11] It is another
matter altogether for us. Argued here is the view that a technological society,
by definition, is a society is at war with the Object. Coexistence is not an
option. The enemy in the crosshairs is the contrary force that contravenes the
performance principle. This force is what entropy is to negentropy, the other

side of a polarity that appears constitutive of a larger order whose existence is regarded as eliminable.

We moderns have a penchant for making an object out of the Object, or a molehill out of a mountain. What works against the technological will to power is perceived as a surmountable challenge and a goad to further action, not a check on the powers of mastery. Now, this way of picturing things may be severely wanting from the lights of the alternative vision emerging here, but it is not utterly delusional. If it were, the folly of our ways would be self-evident to all and there would be no need to mount a counter-argument to the technological. But this is not the case. Most people, it appears, either subscribe wholly to the technological enterprise, consciously or otherwise, or have reservations about aspects of this enterprise while endorsing its overall program. Believers and agnostics outnumber atheists. Few persons are inclined to venture in the direction we currently are heading for the reason that few have misgivings about technology serious enough to prompt a wholesale examination of the ethos that drives technology forward.

To the above can be said the following: It is one thing to subscribe to the view that one ought to work against the force (i.e., nature) that works against us with the aim of carving out for ourselves a more forgiving home, and quite another to believe the world can be remade in a way that eliminates entirely the perversity of things. The difference between the views can be resolved by drawing a distinction between two terms that to date have been used inter-changeably—control and mastery. Consider first the former. Embedded in the word "control" is the notion of adversity. To control something implies acting "against" (*contra*) the "wheel" (*rotulus*) or against the motion of things such as it exists in the absence of a countervailing force. More generously, we could say that in exerting control we accept that our efforts to have the world comply with our designs for it necessarily meet with resistance.

Action understood this way acknowledges the object and suggests that efforts to control nature are merely that, efforts whose success are delimited by the nature of the real. There are many examples of technologies that give us control over the order of things, as just described. A bicycle is an apposite example of a technology whose elegance inheres in the balance between its functionality and its functional limitations. A bike helps its rider traverse distances without losing sight of the fact that doing so comes at the cost of a measure of human exertion. That biking is not a free ride is built into the technology's design and provides a powerful reminder that a cost must be paid for the luxury of expedited movement.

But our interest does not lie with bicycles and other related technologies. Contrivances of this sort do not embody the spirit of technology in its purest form. What does are those technologies and systems of technological order-ing that seek to transform the world in a way that renders it wholly compliant to the human will. Mastery, not control, is their aim. The word "mastery," it should be noted, is linked with notions such as victory and command. Masters by definition have won the battle over command. A master's power over his underlings is total. A similar conceit is shared by the technological vanguard, the architects of the world we inherit, who assume nothing need interfere with the will to mastery. This conceit is as fanciful as the notion that a master's power over a slave is total. We fail to see that the dynamic that underpins a slave revolt is the same one that upends any established system of order. Our blindness in this regard amounts to a denial of the world we are born into. We are world-denying nihilists, as Friedrich Nietzsche might say.[12] Similar to old school Bible-belters, we adhere to an escapist ideology that lays before us a comforting vision of a "promised land" that is our due. Not only the form, but the content of technology's arcadia shares with its more overtly religious counterpart a vision of a reality scrubbed clean of all that makes this world, the immanent world, real. Death and suffering are to be banished. Peace and order alone one day will prevail. Universal prosperity is imminent. As noted, the only substantive difference between the two visions is that we moderns take upon ourselves the God-like task of liberation. And to work toward realizing this end we must believe in the efficacy of the vision that sustains it. The world, we are led to believe, is perfectible: evil can be exorcized.

The belief that the Object is an accidental property of the world, like the color blue is to a ball, authorizes the project to liquidate it. But this is where things get interesting, because if, as argued here, the Object is an essential feature of the real, then the prevailing image of reality remains at loggerheads with the world itself. One would think, as a result, that it should be self-evident that the world does not play by the rules we think it does. Yet there is scant evidence to suggest any such dissonance. The evidence is not there because we moderns have resolved this inner tension in ways that make it seem to have disappeared. This sleight of hand has been affected either by ignoring or misreading the force that frustrates our technological ambitions to gain mastery over the world. With the former, the missionary zeal with which the war against the Other is fought effectively blinds its practitioners to the work expended in the war effort and what this work tells us about the nature of the enemy. On the other hand, with the latter there prevails the "two steps forward, one step back" mentality previously mentioned that, while cognizant of an oppositional force, takes the war as winnable in the long run.

Our disregard of the tension between the world as it is and as we think it is speaks to the power of the technological worldview. We fail to see the Object because objecthood is screened out. Reality for us is refracted through a sensibility that takes the world to be infinitely malleable. Victims of worldview lock-in; we see what we have been conditioned to see and what we have been primed to see is aligned with the technological perspective.

Thinking about technology has led to the conclusion that the technological worldview is problematic on several levels. One adverse consequence is the previously cited disappearance of the subject. The claim was not only that "the human" is becoming redundant but that we are complicit in our undoing. Hardly a hostile takeover, it looks instead as if we have adopted Kevin Kelly's admonition to do what technology wants as modernity's prime directive. How so? How is it that we have become the architects of our own demise? As suggested, a plausible and rather obvious answer is our capture by the technological imperative. Like captors in a version of the Stockholm Syndrome, we have come to identify wholly with the ends of a misguided worldview.

It is time to consolidate this identification, and the best way to do so is to follow Baudrillard's invitation to perform a thought experiment.[13] It involves imagining what we humans look like from technology's perspective. Reversing points of view, Baudrillard asks how humans would appear to a system predicated on the values of functionality, productivity, and efficiency. If your beloved vehicle, for instance, could speak to you in its capacity as a technology, what would it say about its human operator? What would it say about your driving skills relative to its capacities as an operational system? For Baudrillard, the answer is unequivocal. If technology were the judge it would conclude that a human being is an unreliable actor, a veritable "dirty little germ" or "irrational virus."[14] And given our contrary and unreliable nature, if technology had its druthers it would wish for the disappearance of this blight upon the "universe of transparency"[15] and would work toward realizing this end by erasing from us everything that obstructs the vaunted performance principle.

This gedanken experiment is employed by Baudrillard to offset the conventional belief that "we" employ technology to further distinctly human ends. This bias is commonly held for a good reason: it is flattering. We humans make technology, not the reverse, so the creators by definition are in control of their artifacts. In a nominal sense this is clearly the case. But the interesting question here is not who makes what but the sensibility that underlies and guides the making. The simplest way to gain insight into this sensibility is to examine what comes out of the technological tailpipe. Let the real world of technology do the talking. When we do we see that both our artifacts and our relations with them conform to the demands of technology itself. Efficiency rules. So who or what guides the ship? If technology is about mastery it is critical that we get right the nature of the relationship between the master and the mastered. Several decades ago the answer seemed clear, if a little frightening. The consensus then was that technology, like Dr. Frankenstein's monster, had taken on a life of its own and slipped out of strictly human control.[16] Technology, it was said, had become an autonomous force in our lives whose agenda dictated societal development.

The assumption implicit in the technology-out-of-control literature was the continued tension between properly human needs and the ethic of efficiency. What was claimed autonomous is an ethic alien to human ends and the system of means that acts on this ethos. Baudrillard challenges not the autonomy of the technological drive toward mastery but its purported remove from the human. Technology may be autonomous—literally, a law onto itself—but it is not out of human control for the simple reason that we humans identify with technology's "alien" ethic. Technology calls the shots because its vision is our own. The problem today, for techno-progressives at least, is precisely the lingering resistance to the technological ethos. How else to interpret Kevin Kelly's call to do what technology wants but as a plea to surrender *fully* to the cult of efficiency? The problem for technology's boosters is the subset of those dirty little germs who prefer to remain soiled and impoverished creatures. Inefficient, muck and mire humans are thorns in their sides. Nothing threatens the agenda-setting techno-elite more than those unquantified entities who resist the call to self-transcendence and who have the audacity to choose Ithaca over Calypso's offer of immortality.

In an oft-repeated quotation, Arthur C. Clarke once identified advanced technology with magic.[17] For the ranks of the technologically obtuse (i.e., most of us) ours is a magical kingdom whose ways are as enchanting as they are unfathomable. One of the costs of technological advance is that it comes with an impoverishment of our understanding of what Max Weber called the "conditions of life."[18] We moderns, in other words, live off the avails of a system whose operation most of us know next to nothing about. The mismatch between technology's power and our understanding of it contributes greatly to technology's mythologization. Like modern-day animists, our perceived paltriness next to the grandeur of technology leads one to project onto technology a near supernatural aura and power. And if, unlike us humans, technological advance continues apace, the gap between our limited and largely stagnant capacities and technology's ever increasing reach only will widen over time. Less and less will we see ourselves as equal to the products of our ingenuity. What impact this dynamic has and will continue to have on our collective self-image as a species is predictable. If human self-worth is relationally defined, then our perceived status as substandard creatures will become further entrenched into the future, resulting in an expediting of our transition to posthumanism.

Baudrillard foretells this transition to posthumanism. In many important ways, we are already there. Our technological interventions give the lie to Kevin Kelly's claim that technology supplements our humanity in ways that help realize its fulfillment. Quite the reverse, these interventions are better read by Baudrillard as a protracted act of suicide by proxy. The dirty little germs are expendable. Whether "we," through technology, actually succeed in eliminating ourselves from the scene is less important than the will to do so and our acting upon it. And the signs of "progress" are everywhere to be seen. They are revealed collectively in the phenomenon Baudrillard calls "whitewashing,"[19] the business of erasing from the world everything deemed dirty or otherwise reprehensible. Technology, metaphorically speaking, whitewashes reality by cleansing it of those "impurities" that detract from the ideal of perfect functionality. Technology, to extend the analogy, is in the disinfecting business. For Baudrillard, managers, systems analysts, and university researchers, to name a few, are nothing more than glorified hygienists. Each, in its own way, contributes to the larger project that has as its end making a lean, clean, functional machine out of the world. They are, in short, efficiency experts or impersonal technicians whose life goal is the erasure of evil—the Other. This end is approached by the steady removal of everything deemed redundant to the operation at hand. Like plastic surgeons, today's efficiency prettify reality to have it accord with an image of perfect functionality.

While this whitewashing campaign has many fronts, a common tactic in the battle to cleanse reality of evil involves creating barriers that safeguard the integrity of systems, be they biological or social. Crudely framed, we make reality "good" by keeping the "bad" at bay. Technological thinking adopts a prophylactic mindset that links goods such as health and vigor (and proper functioning, in general) with the elimination of the forces that work against these goods. We moderns are conditioned to seek protection from every conceivable threat to our integrity. Politically, this protection comes in the form of rights legislation with its guarantees of personal freedom. Socially, the right to protection is extended in campaigns such as antibullying and child safety programs. These and other social advances imbibe the spirit of technology as much as does the war against cancer and other biomedical interventions. The singular goal is to cleanse reality of those forces that retard the ideal of perfect functionality.

It needs to be stressed that combatting those forces that work at cross-purposes to a system's proper functioning is not at issue. What is questioned here is the nature of the struggle against the forces that oppose the technological will. As noted, because the contemporary ethos admits of no permanent barrier to technological will, there is no perceived counterforce to the drive to mastery. As a consequence, the push to cleanse reality of its imperfections is a radical venture. There is nothing moderate about it. "Reality-fundamentalists"[20] of the technological kind are so convinced of the perfectibility of the real that the effort to integrate reality—to render it a single, unified system—is a total war.

Consider modern medical interventions as a case in point. Baudrillard likes to underscore the fact that the project to protect the human body from all forms of biological aggression is ruinously unrestrained. As he sees it, modern medicine is besieged by a mentality that has it do whatever can be done to barricade the body from attack. The goal is to deny the "enemy" entry, and when and if this tactic fails, to use whatever means available to flush the enemy out. Clearly, the "war room" mindset of modern medical practitioners has met with some success. New drug therapies and vaccines have helped manage the threat of disease and other infirmities. But as it so often happens with technological development, problems with new medical treatments arise because they work too well. The relentlessness of the campaign to protect the human body from all sources of biological contagion has resulted in its hyper-integration as a system. The upshot for Baudrillard is that the body has effectively been shut off from any substantive intercourse with its worldly environment. This maneuver may be ideal from a technological perspective, since it blocks contagions that work to corrupt the system or impair its functioning. But the question to be asked is whether or not the push to create a perfect system out of the human body ideal when viewed from a broader perspective? Are there any negative consequences of our extensive and progressive efforts to shrink-wrap the human body, to isolate the "good" that is functionality from the "evil" that is functionality's underside?

This line of questioning can be rephrased in a way that better integrates our present concern with one that touches more generally on this study's central focus. What, we may ask, would one expect to happen when the human body is treated in a way that adheres to an arguably skewed vision of the order of things? The short answer is that it would not respond as expected. The consequences of this treatment would defy projected outcomes. New and unforeseen complications would arise. Nature, in other words, would avenge herself: the Object would reassert itself. And that is precisely what we see happening, whether Baudrillard is there to tell us or not. For lack of an external threat, systems become incestuously hyperintegrated and produce as a result their own internal virulence or malignancy. The lesson here is that systems need slack to remain viable. When tightened beyond a certain point they turn on themselves. For want of an external threat, systems self-destruct. In Baudrillard's words, "All integrated and hyperintegrated systems . . . tend towards the extreme constituted by immunodeficiency. Seeking to eliminate all external aggression, they secrete their own internal virulence, their own malignant reversibility."[21]

Baudrillard's analysis brings to mind the adage, "The path to hell is paved with good intentions." It is perfectly understandable that we have a vested interest in protecting our bodily integrity. We would not enjoy many of the conveniences of modern day life if were totally indifferent to external and internal threats to our security. The problem, then, lies not in our desire to protect ourselves, but in the extension of the means that have been developed to do so. They are extensive today, so much so that we know, for instance, that the overuse of antibiotics has helped create new contagions largely resistant to available medical treatments. The superbug MRSA is an example of a contagion of this sort, and the fact that it proliferates in hospitals is telling. What it reveals is that the aggressive technological pursuit of a contagion-free environment aids in the emergence of new threats that might not have appeared in its absence.

Baudrillard is struck by the fact that many of today's leading pathologies are immunodeficiency-related. An illness such as cancer, for example, appears to be triggered when defense systems designed to protect their hosts from illness fail to function properly and turn on themselves. It is a disease of our time, he notes, because it is generated by the very success of our medical interventions. The point to be made here is that the human immune system, like any other system, is prone to breakdown should it become unemployed. Pure systems, systems purged of the Object, are fatally unemployed. In working to perfect systems we inadvertently work to set up the conditions of their self-destruction. In promoting them we undo them. This insight appears to be lost on us, or insufficiently appreciated to warrant a reappraisal of contemporary technological tactics. To the contrary, the war against the Object appears undeterred by such setbacks. The war is a total war, as said, and the hubris that underpins it unwavering. Warnings, for instance, that the overuse of antibiotics is leading to their eventual ineffectiveness, with potentially catastrophic results, are left largely unheeded. Self-generated problems of this sort are to be dealt with by finding alternative technological solutions when the need arises. This pedal-to-the-metal mind-set is manic or without restraint. Anything that can be done technologically to improve efficiencies will be done. Any strategy that claims to improve the operationality of a system will be acted upon.

The imbalance at the heart of the technological push to systematize manifests itself throughout our social system. Consider, for example, the phenomenon called globalization in light of the above argument. Globalization is a term that depicts the planet-wide standardization and coordination of ideas and practices, as it pertains to economic, political, and social concerns. Cast in Baudrillardian terms, we can say that the effort to integrate the planet along these lines represents the force of good. Increased systemization produces greater efficiencies, and efficiencies reach their apotheosis with the planet-wide integration and coordination of means. However, in keeping with the impetus behind technological advance, the wholesale effort to dismantle the barriers to a homogenized world order produces its own backlash, or evil. Tom Darby speaks to this same dynamic in his analysis of the tension between civilization and culture, or between the forces of universalism and localism, respectively.[22] The fascistic political upheavals of the twentieth century, he observes, were a reaction to the effects of a perceived growing cosmopolitanism, the chief being the demise of politics, understood in oppositional terms as a contest between friend and foe. In short, fascism was culture's revenge over civilization. It was a revolt against the dismissal of the local and the particular and the wholesale embrace of the universalizing and homogenizing energies of world politics. Arguably, this same dynamic now is being replayed in the twenty-first century with the rise of radical Islam and, more recently, with the resurgence of nationalist politics throughout the West. As with the recent past, the virulence of today's destabilizing political forces results from a disruption of the tension between the need to secure identity, on the one hand, and the demands of efficiency which erode the parochial concerns of cultural life, on the other.

From the vantage point outlined here, then, culture is a virus to body politic of civilization. It represents a special case of the more general techno-logical drive to shield systems from forces that threaten to impair their proper functioning. Baudrillard employs several rhetorical tropes to describe the technological turn toward asepsis, the Boy in the Bubble[23] and Biosphere 2[24] being among them. These references are Baudrillard's answer to George Saunders's "Facility." Their employ alerts us to the growing systematization of all aspects of our social order and the consequences of this development. What we learn from both Baudrillard and Saunders is the *unreasonableness* with which the modern-day forces of reason are unleashed upon the world's disorder. Baudrillard is astonished by the vehemence of the assault against the perversity of things. In part this rush to neutralize evil is due to the very success of past efforts to create a more secure and predictable life for our-selves. If appetite comes with eating, as the saying goes, so does our desire increase for the goods technology brings with every advance in their realiza-tion. The more secure we become the more anxious we are about our present state of security. The more responsive the world is to our interests the less satisfied we are about its current level of responsiveness. Like a positive feedback loop, the "successes" of the technological project breed the kind of anxiety that reinforce our ongoing commitment to the project.

Thinking about technology requires in part that we think about the fit between the nature of our interventionist practices and their patterned out-comes. When we do we realize how poorly the technological world-picture images the reality it purports to describe. Baudrillard contends that this mis-alignment reveals that "the rules of the game" called reality are not what we moderns think they are.[25] As noted, we moderns take on faith the world is constructed in a way that allows for its perfectibility. We then set about perfecting reality by excising from it what does not directly contribute to the ideal of perfect functionality. The targeted enemy goes by the name "evil," or "object," although just as frequently Baudrillard employs "negativity" for the same purpose. Perhaps, though, we should join Czech novelist Milan Kunde-ra and call the reviled other "shit"[26] because the term better captures the intensity of our rejection of a core aspect of the given world. Shit is the counterpoint to the notion that we humans exist in categorical agreement with being, Kundera observes. That is to say, shit is a rude reminder that not all is well with the world. While we know we live in a world where "shit happens," we prefer to ignore what its presence reveals about the constitution of the world. Shit happens, to be sure, but that it does says nothing essential about the world itself. It was noted that we have been encouraged to think this way long before the advent of modernity. Our Christian cultural heritage informed us of the essential goodness of the world and that its present state of corruption is redeemable. So while shit's existence is incontrovertible, for centuries it has been assumed it ought not exist. All this to say that our objection to shit is not that it exists but what its existence appears to signify about the nature of reality. Which is why Kundera rightly calls our objection to shit "metaphysical."[27] It offends because its presence does not fit into our picture of what reality is deemed to be. Like a guest in Procrustes's bed, reality is made to conform to a preset image of itself and shit is not in the picture.

It has been said that every worldview violates or does a disservice to the reality it describes. Yet it also has been said that some worldviews are less representative of reality than others, and that the technological world-picture egregiously misreads the real. In plainest terms, it could be said that modern technology is to reality what ethnic cleansing is to politics. The goal in both instances is to "fix" a problem by eradicating its perceived source. Cleansing reality of shit is technology's mission. And as a subset of reality, excrement-free humans are technology's version of ideal beings: angelic posthumans. For good reason it is difficult to conceive the creatures that populate Kevin Kelly's imagined future shitting.

In working to eradicate from reality everything considered vile and offensive, technology helps create a world of pure positivity—a totalized world, a world of good—not unlike what in the art world is conveyed by the term "kitsch." What is offensive about the vision of tomorrow's engineers is what makes a portrait of a sad clown aesthetically irritating. In both instances what offends is the utter lack of offensiveness. Like bad art, technology is intent on creating a world stripped of every last vestige of unpleasantness. It aims to Disneyfy or deeroticize reality, to make the world a "happy" place where you get what you want, how you want it, and when you want it. In a word, technology works to realize a world that conforms to our understanding of perfection. Through technology, we build a world that works. The desire is to construct a functional world where everything functions with absolute efficiency. We want our politics "fact-based" or cleansed of the rough and tumble of debate. We want our postsecondary institutions to design their educational services in a way that, above all else, meets with student approbation. All things are to follow the fast food model that aspires to deliver a no-surprise experience. Beheld by the technological ideal, we want every experience to be a metaphorical Happy Meal where everything goes our way. Such is the stuff of technology's dream and ours.

The technology delusion is the belief that the world can be rearranged in ways to make it perfectly conformable to human desires and interests. This delusion, we said, stems from a reading of the real that excludes the Object from its world-picture. As a result, it represents a misreading of what it allegedly depicts. Argued here, in contrast, is the view that reality is not amenable to the technological ideal of integrated functionality. This is not to say that reality does not cohere as a whole, but that its wholeness is tensional or suspended between the forces or order and disorder. Baudrillard is a chief spokesperson for this alternative reading. His critique of modernity hinges on its defiance of the technological reading of reality's constitution.

A primary strength of Baudrillard's analysis of technology is its forswearing ethical judgment. As stated previously, in thinking about technology our primary objective is to assess our commitment to the technological project in a way that eschews moral critique. The point, we said, was not to argue that technological advance is either "good" or "bad," but to think about technology in a way that facilitates an understanding of its presumptive end. As a result of such an understanding it may be that a critique of technology is forthcoming, but it is important we underscore that this critique issues from an ontological argument, not a moral one.

So it is vital we repeat that evil for Baudrillard is not evil as traditionally understood: it has no moral dimension, a point he draws attention to by replacing "evil" with "Evil," albeit inconsistently. "Evil has nothing to do with affect," he says pointedly. "It is beyond morality, beyond judgment."[28] Instead, Baudrillard equates the principle of evil with the rule of reversibility. The principle of evil "is simply synonymous with the principle of reversal, with the turns of fate."[29] It is an ontological principle, a special "law of nature" that offsets the forces of order. There is nothing outlandish or radical about Baudrillard's understanding of evil. It is, in fact, an ancient teaching that alerts us to the fragility of goodness, whether we are talking about nature's goodness or our own efforts to make good. The teaching sensitizes us to the "tragic" element in life. Especially, it informs us that any created system that actively eliminates all signs of negativity in hope of attaining perfect functionality is susceptible to a reversal of fortune, to attack by a disintegrating force.

It is important that the term *system* is understood broadly, in this context. Any regimen or organized practice can be pushed to the point where the "good" the practice intends to realize is reversed. This explains, for example, why athletes at the peak of their powers are most susceptible to injury or illness. Training programs that chase perfect conditioning court disaster, as well. This also explains why computer networking systems that seek perfect and total communication, where anything and everything circulates freely in a virtual web of interconnectedness, provoke the emergence of hackers and computer viruses. There is no escaping this fate. Evil is thus *fatal*, in Baudrillard's lexicon.[30] The good of a system is invariably linked with its evil. Because Baudrillard sees good (functionality, order) and evil (chaos, disorder) as necessarily complementary forces, evil for him is not something that can be condemned out of hand. Evil is not "bad." No one, for instance, condemns as "evil" the force called "gravity" should they trip and fall. To be sure, one may suffer the misfortune of falling, but the misfortune has to do with the circumstances that precipitated the fall (i.e., the stone you didn't see on the sidewalk), not with the falling itself. This is why misfortune is rightly associated with the accidental.

Accidents or misfortunes are repairable. We humans can take precautions to avoid accidents, or at least lessen their probability. Evil, in contrast, is ineradicable. There is nothing we humans can do to undo evil, which is why Baudrillard says we moderns at every opportunity convert evil into misfortune. Because we are not equipped to accept the notion of limits that accompany the conceptual framework of a fateful world, we choose not to take a "lucid view" of the reality of evil.[31] We misguidedly assume every problem, every challenge, is "a mere obstacle standing in the way of good."[32]

To be clear, by making a distinction between evil and misfortune Baudrillard is not arguing that things like corruption and disease must be accepted on their own terms, only that to target them for eradication is to misjudge their function in the economy of the real and provoke a backlash more injurious than the problem initially confronted. He asks in this regard, "What is cancer a resistance to, what even worse eventuality is it saving us from?"[33] The same could be said of death. Might death be saving us from eternal life, the ultimate horror? And what, then, of efforts to cheat death by technological means? Are we not revealing in these and myriad other ways that by disrespecting the notion of limits evil is slipping its metaphysical bounds?

Baudrillard's insight into evil is well taken. Many of us instinctively realize that the more complex a system the more things have to go right in order for the system as a whole to go right. Increased complexity therefore brings with it an increased risk of system breakdown. The consequences of collapse increase as complexity increases, too. Take, for instance, the supposition that we humans are arguably the planet's most complex creatures. It doesn't take an expert to realize that our complexity produces outcomes not witnessed in the behaviors and accomplishments of less differentiated beings. We are the best and worst of animals, a fact not unrelated to our constitution. Likewise, the constitution of our social order and its technological substrate is such that we are more highly prone to civilizational catastrophe today than in times past. In many ways, the tremendous explosion in our powers of mastery over the past two centuries has made us more vulnerable than ever to civilizational breakdown. From climate change to pandemics to terrorism, our integrated planet teeters on the edge of disintegration. This paradox is perplexing only if the world is assumed to be perfectible.

There is, however, nothing incongruous about the situation we find ourselves in today when we are disabused of our adopted fantasy-image of reality. It is in the nature of speeding trains to jump the tracks.

We ought to join Baudrillard and ask ourselves how tenable is the view that reality at bottom is a unified state. Is it reasonable to conclude the world is essentially "good," or consonant at least with our perception of good order? Despite popular rhetoric to the contrary, is it not evident that consternation over the perversity of things is fueled by an unrealistic understanding of how the world works? Perhaps this prevailing understanding is worse than unrealistic. Maybe we have the whole picture wrong. Our bias, we said, prejudges reality in a way that portrays it as redeemable. But as far as biases go, the opposite just as well may be asserted. Why ought not the fundamental rule of the game be disorder, in which case the existence of good becomes the exception to the rule requiring explanation? Baudrillard, it turns out, wavers as to whether good or evil is primary. We have seen already that he is partial to the view that evil rules the world. But in a later writing he appears to split the difference between these opposed prejudices by suggesting they coexist. Rather than pick sides, Baudrillard highlights the dualist ontology of good and evil, arguing that it better accounts for worldly dynamics than any monist alternative. As he puts it, "I regard duality as the true source of all energy, without, however, passing any verdict on which of the two principles—good or evil—has primacy."[34]

It is this duality that constitutes for Baudrillard the previously cited "rules of the game" that account for the world's dynamic tension. The world's energy draws from the interplay between good and evil. Both the forces of order and disorder are integral to the being of the real. Efforts to reorder the world in ways that conform to our understanding of good order alone are therefore bound to be frustrated. The rules ensure that our designs for order are always challenged or met with a contrary show of force. This is how the world's indifference is manifested. Reality's logic remains resolutely dualist, in total disregard of our ambitions to create a world founded on the unitary principle of order. So a Baudrillardian response to Kevin Kelly's rhapsodizing about technology would be simple enough. One's blathering about the virtues of technology will not alter its fate.

NOTES

1. While I have appropriated Jean Baudrillard's oeuvre for my purposes here, I believe I have done so in the spirit of his ambition to comprehend and critique the forces at play in the contemporary world.

2. Baudrillard lays the matter out succinctly with this claim: "This is the miracle: that a fragment of the world, human consciousness, arrogates to itself the privilege of being its mirror. But this will never produce an objective truth, since the mirror is part of the object it reflects." And then he concludes, "The task of philosophy is to unmask this illusion of objective

reality—a trap that is, in a sense, laid for us by nature." See Baudrillard's "On the World in its Profound Illusoriness," in *The Intelligence of Evil: Or the Lucidity Pact*, tr. Chris Turner (London: Bloomsbury, 2013), especially 29–31.

3. Jean Baudrillard, "Forget Artaud," in *The Conspiracy of Art*, ed. Sylvere Lotringer, tr. Ames Hodges (New York: Semiotext(e), 2005), 221.

4. Ibid., 163.

5. Baudrillard devotes an entire book to the theme of a perfect crime. See *The Perfect Crime*, tr. Chris Turner (London: Verso, 2008). See also his "The Perfect Crime," in *Passwords*, tr. Chris Turner (London: Verso, 2003).

6. Baudrillard, *Passwords*, 66.

7. Umberto Eco, *Foucault's Pendulum*, tr. William Weaver (New York: Harcourt Brace Jovanovich, 1989), 95.

8. Baudrillard, *Conspiracy of Art*, 170.

9. Ibid., 234.

10. I realize I may be taking liberties with Baudrillard's reading of the object by linking it with evil, but it strikes me that as an oppositional force to the good, evil objects, or is thrown against, the good. The Object, thus interpreted, denotes a metaphysical principle.

11. E. R. Dodds, *The Greeks and the Irrational* (Berkeley: University of California Press, 1951), 49.

12. Friedrich Nietzsche, in aphorism 56 of *Beyond Good and Evil*, equates the world-denying worldview with the likes of Schopenhauer's philosophy and "Asiatic" philosophies such as Buddhism. This nay-saying orientation is juxtaposed against the "ideal of the most high-spirited, alive and world-affirming human being who has . . . come to terms and learned to get along with whatever was and is." See Friedrich Nietzsche, "Beyond Good and Evil," in *The Basic Writings of Friedrich Nietzsche*, tr. Walter Kaufmann (New York: The Modern Library, 1992), 258.

13. The invitation is not literal but implied in Baudrillard's analysis of "the extermination of man." See his "Prophylaxis and Virulence," in *The Transparency of Evil: Essays on Extreme Phenomena*, tr. James Benedict (London: Verso, 1993).

14. Ibid., 68.

15. Ibid., 68.

16. I am thinking here of the impact of Langdon Winner's influential text, *Autonomous Technology: Technics-out-of-Control as a Theme in Political Thought* (Cambridge, MA: The MIT Press, 1978).

17. The full quotation reads: "Any sufficiently advanced technology is indistinguishable from magic." It is found in Arthur C. Clarke's *Profiles of the Future: An Inquiry Into the Limits of the Possible* (New York: Holt, Rinehart, and Winston, 1984).

18. Max Weber, "Science as a Vocation," in *From Max Weber: Essays in Sociology*, eds. Hans H. Gerth and C. Wright Mills (New York: Oxford University Press, 1946), 139.

19. Baudrillard focuses on the theme of whitewashing in "Operational Whitewash," from his *The Transparency of Evil*.

20. Baudrillard, *The Intelligence of Evil*, 23.

21. Baudrillard, *The Transparency of Evil*, 69.

22. See Tom Darby's "On Odysseys," in *Sojourns in the Western Twilight: Essays in Honor of Tom Darby*, eds. Robert C. Sibley and Janice Freamo (Eastern Townships, Quebec: Fermentation Press, 2016).

23. Baudrillard, *The Transparency of Evil*, 68.

24. See Baudrillard's *The Illusion of the End*, tr. Chris Turner (Stanford, CA: Stanford University Press, 1994), 87.

25. The phrase "rules of the game" appears in multiple places throughout Baudrillard's writings. In *The Intelligence of Evil*, however, Baudrillard sums up nicely what he means by the expression by linking it with duality, reversibility, and evil. "It is duality that fractures Integral Reality, that smashes every unitary or totalitarian system by emptiness, crashes, viruses, or terrorism." He then concludes, "There is no point deploring this—nor exalting it for that matter. These are quite simply the rules of the game. Everything that seeks to infringe these rules, to restore a universal order, is a fraud" (145–46).

26. See Milan Kundera, *The Unbearable Lightness of Being*, tr. Michael Henry Heim (New York: Harper Perennial Modern Classics, 2005), 131–33.

27. Ibid., 131.

28. Jean Baudrillard, *The Agony of Power*, tr. Ames Hodges (Los Angeles: Semiotext(e), 2010), 119.

29. Baudrillard, *The Transparency of Evil*, 73.

30. Baudrillard, *The Intelligence of Evil*, 107.

31. Ibid., 111.

32. Ibid., 107.

33. Baudrillard, *The Transparency of Evil*, 74.

34. Baudrillard, *Passwords*, 86–87.

Chapter Five

Hall of Mirrors

"The rules of the game": what an evocative way to convey the order of things and how refreshingly at odds with the contemporary preference. We moderns are nothing if not serious about reality. Enormous resources today are expended plumbing its depths, getting to the bottom of the real with our measuring, analytics, and evaluations, all with the expectation that the knowledge we retrieve will lead to equally serious benefits for humankind. There is not a whiff of levity in the whole enterprise, no sense of it being an elaborate game the outcome of which, like all contests, cannot be determined in advance. Friedrich Nietzsche, over a century and a half ago, understood well the grimness of modern rationality and offered his "gay science" as an antidote.[1]

Richard Feynman, the late Nobel Prize–winning physicist, was an American eccentric with a compulsively playful personality.[2] Yet he was deadly serious when it came to matters pertaining to truth claims. In a film devoted to his life and works, Feynman recounts a story where his father, pointing to a warbler, begins listing the names of the bird in various languages.[3] Then, as Feynman relates, his father concluded, "You can know the name of the bird in all the languages of the world, but when you are finished, you'll know absolutely nothing whatever about the bird. You'll only know about humans in different places, and what they call the bird. So let's look at the bird and see what it's doing—that's what counts." That's what counts. Some things count and others do not. Implied here is that there is a path (a "way after," from the Greek *methodos*) that leads to true understanding and cul-de-sacs that lead nowhere.

But what counts for what counts? On what purely rational grounds can one argue that identifying a bird as belonging to a particular species of animal, by attaching a name to it, tells us absolutely nothing about the bird? To say it does not implies that language supplies us with a mere image of reality in the symbolic form of an utterance. It suggests a sharp distinction between the real and its representation, and that there exists a direct path to the real, one that bypasses the realm of symbolism and representation.

The assertion that there is *a* path to the truth—the scientific path—is deeply embedded in contemporary Western culture. It is replayed in the distinction between myth and reason. Myths are just stories, after all. We moderns are superior to our primitive forbears because we understand myths to be what they are in fact, mere tales, unlike those naive premoderns who actually believed their stories to be true. This modern reading, we fail to see, reflects an interpretive bias toward the world and our relationship to it that is contestable.

We can turn to Plato for insight into how it might be possible to reimagine the world in a way that opens up what counts as knowledge. Along with Socrates, the ancient Athenian is rightly credited with the founding of the Western philosophical tradition: the birth of reason (*logos*). Written almost five hundred years before the New Testament, one is struck when reading Plato's dialogues by how contemporary they sound relative to the Gospels. The dialogues bristle with an interrogative spirit. Absolutely no idea or position is taken on faith. Everything is subjected to rational scrutiny. Very modern indeed. Yet the analyses are presented in dialogic form, as literature. Plato's intellectual discourses have a story-like quality to them. They are imaginative portrayals of real characters (or character types) in real settings talking about issues of real concern to the discussants. So built into the very form of Plato's dialogues is a blurring of the lines separating truth from fiction. Clearly Plato had no problem with employing art in the service of truth. More interesting still is the fact that Plato frequently incorporated myth into his reasoned discussions.[4] Many a time, in the midst of a heated debate, an interlocutor will interject something to this effect: "Let me tell you about a story I once heard."

What can be made of Plato's seeming ambivalence toward "the truth"? Why was he so at ease slipping between *mythos* and *logos*? Did he not fully comprehend the distinction between the two? The typical modern response is to think he did not. Reading Plato with a historical sensibility, the contention is that because Plato lived in a time of transition between the archaic and the modern eras his preference for rational thought over mythological thinking was not categorical. But is it possible to imagine what Plato might have been thinking about the business of thinking without the supposed benefit of hindsight? Is it conceivable that Plato understood fully the nature of *mythos* and *logos,* and concluded the two modes of revelation are not diametrically opposed? There are reasons to believe Plato was cognizant that all forms of rhetoric reside within the realm of symbolic representation, that philosophical analysis, as much as storytelling, remains metaphorical at bottom. If so, then the difference between the two becomes primarily a matter of form. A good story can tell us something important and true about the world, but it does so in a more compact or less fully articulated (and thus self-conscious) form than an analytic treatise on the same subject. In neither case, however, and this is the essential point, is the account fully adequate to the reality it attempts to articulate.

I hope the reason for this digression is becoming manifest. What counts for what counts (for knowledge) for Plato is not what counts for someone like Feynman because they adhere to two radically different perceptions regarding what constitutes true insight into the nature of things. Feynman places a burden on scientific knowledge that Plato never placed on his attempt to rationally comprehend the world. For the former the scientific method clears a path to understanding things as they *really* are. No such path exists for Plato. Even when Plato gets metaphysical, as he does with his discussion of the Theory of Forms, he realizes the theory remains an image of reality, a depiction of reality, and therefore *not* the real itself. So even if the theory in question were a true depiction, as a perceptual account *of* reality a distinction is upheld between the image and the imaged.

That knowledge is a mode of perception key to understanding both Baudrillard and the alternative to the technological worldview advanced here. It has been argued that Plato understands our understanding of reality is necessarily allusive. Every effort to capture reality conceptually falls short of the mark of identifying with the territory called "reality." Pushing this reading of Plato to the limit, we could say that his corpus consists of an extended effort aimed at illustrating the folly of assuming that any articulation of the truth constitutes the Truth. In a world where many lay claim to having discovered the Final Word, or The Path to it, Plato reveals there is no such word, no such path. The object of our desire—Reality, the really real—is inaccessible to us.

That the truth of the world remains forever elusive suggests the world toys with us in some mysterious way. We are lured by an unfathomable impulse to search for something that cannot be found. It is confounding that implanted in us is an itch that cannot be scratched and equally perplexing the world that makes this our fate. For Baudrillard there is no accounting for the human predicament or the world that gives rise to it. The world is what it is, and we either accept that the world plays with us through its reversals and irresolvable dualities or we wish it away. Regardless of the choice we make, for Baudrillard we remain suspended in a kind of cosmic tug of war over which we do not have ultimate control. There are no "laws" in this universe prescribing preset and predictable patterns of movement and outcome, only rules, as in rules of thumb. That is to say, while the world appears to operate according to a set of underlying principles, its adherence to these principles is not absolute. There is an in-built latitude and unpredictability to the order of things that keeps it from being strictly ordered. So if reality possesses an order, it is decidedly nonlinear.[5]

When it is assumed the world functions as a law-abiding system, the door opens to the possibility of understanding its laws in ways that permit their utilization to our advantage. In calling reality a game, however, Baudrillard seeks to have us understand the inherent limits of any effort to comprehend and master reality. Reality always gets the last laugh in the push and pull of our mutual exchanges. We seldom win, and if we do never in the way we intend to win.

It may seem incongruous to those who identify Baudrillard with a brand of radical intellectualism that he takes his cues from nature. Yet he does. Given, he reasons, that the world plays with us, the only fitting response is to return the favor and toy with it. Baudrillard answers the allegorical nature of our ideational relationship with reality with appropriately figurative attempts at articulating reality's contours. The purpose of Baudrillard's prose is to adjust word and world to have one resonate with the other, not to explain the unexplainable. The informality of an expression such as "the rules of the game" is therefore intentional. Baudrillard only can allude to the constitution of reality given his understanding of it. The expression remains a conveyance, as does Plato's Theory of Forms. This is the best that can be done. We moderns think differently, of course, as we work diligently to construct a model of reality that gives a full account of the essence of the real. This is our folly, as Baudrillard sees it.

Much of the preceding analysis of the effects of technology is a riff on what happens when we refuse to play footsy with reality. Strange and unexpected things happen when we fail to understand that the actor is always, at the same time, the acted upon and that what acts upon us is largely indifferent to properly "human" concerns. We have focused to date on one form of unforeseen consequence, namely, the recurrence of technology-induced blowback and have addressed its import. But there is another type of technology-related payback that needs discussing. This has to do with perceptions of reality. Specifically, it relates to the growing confusion and indifference towards the meaning of the real.

The connection between technology and "the reality problematic" cannot be properly addressed outside a discussion of simulation, a theme of central importance to Baudrillard. We have discussed at some length already the question concerning the nature of the real because this question was deemed intimately tied to technology as a worldview. Baudrillard, we observed, believes there are no grounds for drawing a clear line between the real and the unreal, or the authentic and the inauthentic, because the "real" world for us is out of bounds. On this account, thoughts about things are simply that, thoughts *about* things, not the things themselves. So the realm of ideas (and ideals) and the real world to which ideas are directed are kept distinct. One of the strange and unexpected consequences of our rejecting reality's illusory character is the conflation of the real with its image. Ironically, in our effort to "get real"—to understand the world as it "really is" and to reorganize the world according to this true understanding—our grasp of what constitutes reality begins to loosen.

Baudrillard's analysis of simulation highlights the perverse consequences of living in an age obsessed with reality—the age of technology. As noted, what makes our age properly technological is "the perpetrating on the world of an unlimited operational project."[6] We moderns work to perfect reality, and we do so by having ideas pertaining to what ought to be serve as models for what will be. In the final analysis, the fusing of the ideal and the real—of thought and action—is what technology is about, and lies at the heart of what Baudrillard means by simulation. Ours is a world where ideas rule. More and more, everything we engage and interact with is a product of human ingenuity, conceived in accordance with the principles of efficient design. Renovating reality in this manner is possible because the means of remodeling are deemed adequate to the task of remaking the real in a way that conforms to our understanding of proper functioning. Our certainty, in this regard, legitimates the reconstruction project. And what has emerged as a consequence is a simulated reality. Baudrillard is a frontrunner in calling attention to the fact that we have reached a level of technological sophistication where the "real world"—the world of everyday experience—is almost always a reconstructed reality: the real has been replaced by its objectless twin.

Employing an image or model of the real as a basis for reconstructing the real is what we moderns do for a living. We model everything. We construct business models, management models, consumer behavior models, pedagogical models, governance models and the like, and employ them as a means of increasing the efficiency of desired outcomes, whatever shape they may take. Modeling, then, is not merely a component of the noun *technology*: it *is* technology in action. The technological mind is a modeling mind, and we inhabit a world remade to accord with this mind's image of good order.

Plato never would have conceded that the world exists to be reconstructed to align with an idealized image of reality. His imagined republic, his fabled "city in speech," was no blueprint for political action. If anything, it was a veiled critique of the mentality that assumes humanity would be better off living in a world that corresponds to an idealized (or objectless) image of the real. A perfectly functional state is antinature and thus inhuman, and Plato's respect for the limits of rational ordering speaks to his humanism. We moderns see things differently. If functionality is antinature, then tough luck for nature and for us as natural beings. A reworked or simulated reality that works is preferable to inhabiting a world that accepts the limits of functionality.

To understand better what Baudrillard means by simulation we first must consider its everyday meaning. A standard dictionary definition connotes simulation with imitation or enactment. A simulation, accordingly, is a reproduction of one sort or another. As such, simulation is linked with artifice. The artificiality of simulations immediately suggests its antithesis and the tension between the two. There exists on the one hand a category of real or authentic events (either natural or humanly created), and on the other a category of feigned or counterfeited ones. We live in a world where there are real pearls and faux pearls, real Italian bread and Italian-style bread, and so forth.

To understand how the distinction between real and fake is being lost today, we have to take a closer look at what Baudrillard specifically means by the term simulation. Luckily, he gives us plenty of help to this end. As already suggested, to simulate is to represent one thing by means of another. Fake pearls, to return to our example, are crafted to resemble real ones. They represent the real thing, and if the representation is faithful they resemble closely the reality they copy. This raises an important point about simulation Baudrillard underscores. "To dissimulate [to lie]," he observes, "is to feign not to have what one has. To simulate is to feign to have what one hasn't."[7] So to lie about being sick, for instance, is merely to pretend not be healthy, to say perhaps that one has a stomachache when one does not. To simulate sickness, on the other hand, is to invoke the symptoms of sickness or to induce artificially the actual experience of stomach pain. Simulated pain therefore is both real, insofar as is actually experienced, and unreal to the extent it is artificially induced. Simulation, on this account, straddles the dividing line between truth and falsity, or between the natural and the artificial.

A return to an examination of flight technologies might help better illustrate simulation's ambiguous relation to the real. We can begin by referencing a gag, one of many, that plays with the thin line separating reality from artifice. It has an airline flight attendant ask her passengers, in response to the jet's mechanical failure, if anybody knows how to play Microsoft's Flight Simulator.[8] While it ostensibly pokes fun at the notion that flying a real jet and a virtual one are almost interchangeable experiences, the joke is on us because for all practical concerns they are. Simulators at flight training schools are elaborate pieces of machinery that faithfully replicate the experience of flying real aircraft to the point that trainees have been known to suffer heart attacks as a result of trying to handle a virtual flameout. On the other side of the ledger, piloting a "real" jetliner these days is strikingly similar to flying a simulator, save the odds of surviving a crash in a real plane. It was noted previously that the routine task of flying such aircraft has little to do with old-school piloting. The glass cockpit experience amounts to monitoring the monitors of an essentially self-guiding system. So accustomed are pilots to the rote work of managing flight systems that when on occasion circumstances require them to fly manually, they sometimes are caught unprepared.[9] "Too bad there wasn't a pilot onboard" is both comic and tragic commentary on the state of modern aviation.

So, in this strange world of ours, simulations of flight are largely inter-changeable with the real thing and the real thing is almost identical to a simulation. Fake flying and real flying are largely indistinguishable experiences, rendering the distinction between "fake" from "real" meaningless in this context. Stated more abstractly, what we are witnessing here is the merging of image or representation and the thing imaged or represented. The gap between the virtually real and the real is disappearing.

There are many other examples of this technology-induced disappearing act. Take, for instance, the modern practice of "molecular modeling,"[10] where computational techniques encoded in computer software programs allow chemists, biologists, and material science experts to replicate the behavior of molecules. The benefit of working in so-called "dry" labs with virtual versus real molecules is obvious and in keeping with the primary value of technology: efficiency. Real (versus computational) scientific experimentation is expensive in terms of cost and time. It is much more efficient, and hence more productive, to model reality prior to performing those laboratory experiments that have been determined virtually as potentially promising.

So what is the significance of this kind of modeling? It tells us that by having the virtual modeling of chemical reactions *precede* the actual reactions, the "real" reaction is a product of artifice. This is the reverse of our everyday sense of things, where we assume reality precedes images or reproductions of the real. Every time, for instance, we witness an event in our lives and memorialize it by reproducing the event in image form we reinforce the notion that images are necessarily images *of* reality. But what if, as in the case above, reality's very existence—the blood pressure pill you took this morning or that flight you boarded later in the day, for example—was conjured in the circuitry of a computer? If the "real" is a replication of an image without an original, how can we think any longer of the real as a self-subsisting thing, as standing apart from the human domain?

Test marketing is another common practice that produces the same result. Private businesses and public sector organizations routinely hire research firms to test-market the saleability of consumables, whether they be ideas, products, or public policy proposals. The information gathered is then folded into the crafting and delivering of these consumables. This practice is performed in the name of efficiency. It is never good to waste resources producing something that meets with public disinterest or disapproval. And this practice produces the same result as mentioned above with respect to aircraft technology. What comes out of the production pipe is a real copy of a pre-tested or virtual reality: a simulation.

Simulations or simulacra are everywhere. Theme parks and restaurants, museums and historical sites, reality TV, these are some of the more overt examples of simulation. But to speak of simulation by referencing alone concrete instances of the phenomenon is to mistake the forest for the trees. Baudrillard has noted, famously, that "Disneyland is presented as imaginary in order to make us believe that the rest is real."[11] His point in obfuscating this distinction is to make known that we inhabit instead a simulatory *universe*. For Baudrillard it is a mistake to assume the "real" world is populated with isolated pockets of simulation. This error is committed because we like to believe the "reality principle" remains intact, despite our penchant for technological intervention. But such is not the case. Baudrillard argues our intercourse with "reality" is always already mediated in a way that has reality effectively disappear. Simulation is everywhere, which is to say our "world" is thoroughly artificial.

Let us return to our discussion of George Saunders's "Jon" to understand better Baudrillard's seemingly extravagant claim regarding simulation. We noted that Jon's world reflects our own consumerist society. It parodies ours insofar as it presents a life lived entirely within the confines of a self-designed consumerist ecosystem. The satirical intent is accentuated by the fact that Jon's confinement is literal: Jon lives in the bubble called the Facility. Everything Jon comes into contact within this "world" within the world is contrived by definition. And further, it is contrived for the singular purpose of bolstering the efficiency of the consumerist society he inhabits.

Jon, like his cohorts in the Facility, is characterized as a real person possessed of real human emotions, interests, and needs. He is, among other things, a sexual being, a social being, and a person concerned with his own material welfare. All of these yearnings were attended to in the Facility. We noted that while the Facility was a consumer satisfaction machine, it was also a mechanism that secured the satisfaction of its functionaries. Jon's sexual frustrations, we saw, were responded to as a computer technician might react to a software glitch, with dispatch. Likewise, the need for purpose and direction in life was met by supplying him with a function of distinction. Jon's all-too-human need for emotional security and stability also was ministered to, in typical fashion, by artificial means. All together, Jon's life was simultaneously real and unreal: it was a simulation.

Jon's contrived existence was satisfying, at least initially. The contriving aimed to produce self-satisfaction and largely succeeded. Technological advance seeks the same overall end. Your iPhone is just the latest iteration of a technology that wants to do what technology more generally wants to do—make you happy. It wants to help you find friends, listen to the kind of music you like, and navigate through the maze of consumer options it lays before you. But it is precisely as an aid to self-satisfaction that technology holds the greatest danger, for Baudrillard. What is most commonly valued about technology, its operational capacity, is also what makes technology most problematic.

Consider again in this regard Jon's life in the Facility. Before the fall, Jon thought himself empowered. Life went his way. The problem is that his self-empowered existence was artificially engendered and sustained. Jon might have thought himself a big shot but he was just a two-bit player in a game he knew nothing about. In this sense his "real" life was a fake. This fakeness, however, is evident to the reader, not Jon, the former being positioned outside Saunders's imagined reality. From the inside, things look differently. The fakeness of Jon's real life is not a datum of consciousness for him. On the contrary, the artificiality of his existence has to be withheld from him in order for the illusion of realness to be sustained.

If, as argued, Jon's life is a parody of our own, then we must read our technological society as we read Saunders's "Jon." Thinking about technology means engaging in this kind of interpretive exercise. It entails a self-distancing, taking an outsider's view of life on the inside of the technological dynamo. It involves asking ourselves to what extent our "real" lives resemble Jon's, and how our embeddedness in the technological matrix affects both its evolution and our understanding of it.

To review, Baudrillard contends modern technological society is so thoroughly reconstituted that our simulated environment is taken as real. Whether through ignorance or indifference, our retrofitted reality is treated as real, which is tantamount to saying that reality has disappeared. So in this surreal world of ours, images or recreations are real and reality is an image of itself. Everything becomes its own icon, including natural things. It becomes impossible, in this context, to disentangle your "real world" experiences from recreated ones since your "real" experiences are almost always the product of artifice. Immersed in an image-world, we conflate the image with reality, with the world itself.

A perfect instance of this conflation is Lascaux II.[12] The Lascaux Cave, as well known, was an important archaeological find that served as home for a stunning array of Paleolithic art. However, due to concerns over the impact of human visitation, the cave eventually was closed to public viewing and replaced, some twenty years later, with Lascaux II. Situated near the mouth of the cave, Lascaux II replicates the cave art and its setting faithfully. It stands in for the original cave—the renamed Lascaux I. So when tourists visit "Lascaux" today, they experience its simulation. If a person were to claim to have seen Lascaux, she will have seen its replica. Thus the real Lascaux effectively has disappeared. Reality is image. Lascaux II is the trace of the real that cancels the real it replaces.

For Baudrillard the entire planet is fast becoming a Lascaux II. The technological ethos seeks to make of the world an unlimited operational project, we noted.[13] Lascaux II, or any other instance of simulation, is representative of a civilizational commitment to whitewash or operationalize reality. What Baudrillard brings to our attention in speaking to simulation is its connection with the general disappearance of "objective reality" or "reality related to meaning and representation."[14] To explain this linkage Baudrillard's usage of the expression "objective reality" first must be clarified. Clearly, a good measure of humor is folded into its employment. The *real* world, Baudrillard informs us, is not what we moderns think it is—the world of things and facts. Objective reality for him does not pertain to what can be brought within the ambit of rational understanding and control. The really real, in other words, has nothing to do with the preferred technological approach to reality. On the contrary, as stated earlier, reality is always an imagined reality, a world invested with meaning. And importantly, reality is meaning related precisely because meaning is illusive. That meaning is not self-evident means meaning is necessarily imagined. So-called objective reality, then, is an imagined reality. Our technological society is grounded in an imagined reality, too, which ironically rejects the centrality of meaning and imagination.

In linking illusion with meaning and imagination Baudrillard attempts to upend the notion that reality can be conceived independent of its imaginative dimension. There is for him no factual reality, per se. To assume such a reality exists is to posit it may be known in its totality. It is to assume that nothing exists beyond the limits imposed upon reality in its actualization by technological means. That is why Baudrillard asserts that a simulated universe "has no imaginary."[15] Reality disappears "metaphysically," he concludes, when its principle dies, when we no longer experience the world as transcending its actualization.

To explicate further the implications of Baudrillard's claims regarding simulation, technology, and the death of the real, we can return yet again to Saunders's "Jon." Ostensibly, what was most alarming about Jon's existence in the Facility was not his unfreedom but the impact of his conditioning on his imaginative life. In effect, he did not have one. Jon's nightmare was his living entirely within the prescribed imaginary confines of the social order he inhabited. His hell was the hell of the same. Differently put, reality (as Baudrillard conceives it) was dead to Jon. Its principle was moribund. Any innate capacity he might have possessed to imagine a world beyond the Facility was counteracted by a program that sought its liquidation.

Life in the Facility was engineered so as to have no one want for anything. It was designed to eradicate *eros*. Effacing the real, which amounts to actualizing the real, was the route to this end. Producing deeroticized souls required aligning the real and the imaginary, or collapsing the tension between what is and what might be. The same dynamic is operative here and now in the "real" contemporary world. The ideological function of technology is to have us hitch our hopes and ambitions to the promise of technology, to want what technology wants. To the extent this end is realized and consolidated, we lose, Baudrillard says, "the passion for reality and the passion for truth."[16]

The fact Baudrillard sees technology as connected with a growing indifference to the real is informative. As noted, all passionate or erotic attachments presuppose lack. A passion for reality or truth presumes some kind of problematic relation to them. The problem with technology is that it works to undermine our problematic attachment to the world. Both as a worldview and a practice, technology arranges the world so as not to experience it in opposition to human want.

The imaginary, in effect, works to keep the real in bounds. To imagine what is not is to realize that what is is not everything. It is the gap separating worldview from world, image from imaged, actual from imaginary, signifier from signifier, which Baudrillard wishes to reestablish in opposition to the forces of closure. His universe, we said, is dualist. He conceives the world as tensional and we humans as caught in the erotic interplay between dualities.

Baudrillard's take on film and photography speaks to his dualist proclivities. In Baudrillard's estimation, art at its best mimics or represents the erotic encounter between perceiver and perceived. If reality is an illusion, he suggests, the imaging of an illusory reality ought to reproduce its illusory character. The problem, for Baudrillard, is that the art of imaging tends to move in the opposite direction, toward realism. Speaking of film, for instance, he argues that increases in "cinematographic efficiency" necessarily result in the withdrawal of illusion.[17] In other words, for Baudrillard film gets less real as the image more faithfully reproduces the real. He wishes the image to remain an image of reality, to allude to an allusive reality, while the technological zeitgeist pushes for representations that, in their pursuit of verisimilitude, capture reality and ultimately substitute for it. High-definition imagery therefore is unreal because it fails to do justice to the reality it purportedly portrays. Such imagery only reinforces the illusion that the essence of the real is capturable. The best art resists the temptation to render reality real. Like our perception of everyday reality, art ought to undersignify. This stands in contrast to a cultural milieu that habitually does violence to the image by overrepresenting it.[18]

Not only is realism a misguided approach to aesthethic creation, for Baudrillard it is fraught with real world consequences of considerable import. A primary problem pertains to the fact that once reality is perceived capturable, once there is no perceived check on realizing the real, the realizing process knows no end. Reality gets hyperrealized. Recreated reality gets hyperrealized by taking on the character of an ideal form of the real. If simulations are realer than real, it is because they are models of reality purged of imperfection. In the realm of popular music, for example, the drive toward hyperrealization is manifested in the ongoing trend for music to disappear behind its own special effects. Both in its production (via pitch-control devices, audio software programs, etc.) and consumption, popular music almost entirely has become its own ideal image.

The evolution of modern advertising exhibits the same trend that has the image occlude its referent. The logic of the history of modern advertising is the logic of the hyperrealizing of advertising imagery. Once, for example, promoting motor vehicles required that advertisers disseminate vehicle-related imagery and information. The talk then was of smooth rides, horsepower, and great handling. While advertising has not entirely dropped concerns of this sort, increasingly the focus of campaign advertising today is on the image to the exclusion of the imaged. The advent of lifestyle advertising marks the transition to imagery liberated from content. Advertisers sell images and ideas, not things. It is the image of a way of life a particular product is crafted to signify that sells. In general terms, this means consumers purchase items less to satisfy tangible needs than to reflect the kind of person they think they are and the values they assume they represent. The act of consumption involves matching self-image with product image or brand.

Advertising imagery is hyperrealized in the same way Lascaux II constitutes a realer than real image of what it ostensibly represents. There is one important difference, however. Simulations such as Lascaux II remain bound to their referents in a way that hyperrealized advertising imagery does not. By linking brand with self-image, lifestyle advertising makes it possible for almost any image to be attached to any product or referent. No one would think it inappropriate, for instance, if cockatiels or balloons were the visual focus of an advertising campaign for jeans. To the contrary, juxtapositions of this sort likely would be praised as creative should they succeed in increasing sales. The upshot is that no image is potentially exempt from being associated with any referent. Anything can represent anything else.

For Baudrillard, the liberation of the signifier (or image) from the signified (or imaged) means the signifier can refer to anything and everything. This means in turn that the boundaries that once distinguished domains such as politics, aesthetics, economics, and sexuality have all but entirely dissolved. And when these boundaries disappear everything comes to be seen as equally political, aesthetic, sexual, and so on. This is a problem, Baudrillard contends, because when everything is political, nothing is political; when everything is sexual, nothing is sexual; when art is everywhere, nothing holds aesthetic power anymore.

In saying this, Baudrillard is not suggesting that in the past there were sharp and rigid delineations between these significatory realms. Signs always have had the capacity to cross boundaries to an extent. However, in the old regime, the reality principle acted as a brake to the unfettered circulation of signs. This principle was the force that kept signs tethered to the real world of things and their usage, and today it has all but disappeared. For Baudrillard, the realm of image now constitutes its own domain, its own reality. It has overtaken the things that signifiers were intended to represent and we, as consumers, have become consumers of signs.

The radical liberation of signifiers has another important consequence. The free commingling of signifiers allows for an increase in the rate of both their production and circulation, resulting in an "epidemic of value" and meaning.[19] For Baudrillard, it is as if the regulatory mechanism that once governed the sign universe has collapsed, resulting in the sign-world's uncontrolled growth. The image- or sign-world, unlike the natural order, has no checks on its exponential growth, no barrier that might contain the continued eruption of the hyperreal, or reality's proxy—the world of signs and simulations. Like a cancer, signs metastasize and swirl about at light speed. So the hyperreal sign-world is by definition a viral realm. It is spinning and spewing out of control like uncapped oil well. This excess, Baudrillard is convinced, leads to the system cracking up.

The system implodes through processes of reversal. If, we have seen, overly positive systems of control are vulnerable to blowback, so too is excessive signification susceptible to its own form of reversal. Just as traveling at a high rate of speed can yield the uncanny sensation of stillness, so, Baudrillard says: "Nothing [is] more unreal than the accumulation of facts." In other words, the perceived reality of a situation is diminished, not amplified, if too much reality is evidenced. One death is a tragedy, a million deaths a statistic, as the saying goes. The principle at work here says that too much of anything undoes that thing, reverses its effect. So an excess of information results in the "undecidability of facts and confusion of minds," an excess of death in moral indifference, an excess of security in immune system failure.[20]

We do not need Baudrillard to tell us that contemporary culture is image saturated. And we do not need him to understand that the flood of images that washes over us leads to a kind of psychic numbness where inputs stimulate and excite but rarely grab hold or cut deep. Information overload likewise leads to a kind of intellectual numbness where managing masses of data calls for the development of skills that militate against the capacity for deep reading, once seen as a mark of erudition. "Doesn't information kill education?" Baudrillard asks rhetorically.[21] Absolutely. He would emphatically agree with Nicholas Carr that Google is making us stupid.[22] The information space we ply is suffused with signs to a degree that it forces upon us a posture of indifference, where scanning is a virtue. This adaptive strategy may be effective given the media environment we inhabit today, but it works against the cultivation of forms of sustained thought, which helps explains, for example, why it is increasingly difficult to get students in universities to appreciate, let alone engage in, the kind of critical thinking educators and administrators claim they are interested in promoting.

Academics, especially in the arts, often are subjected to ridicule for their excessive theorizing. Admittedly, this derision is not always unfounded. However, what goes unrecognized by many of the critics of the "unproductive" intellectual class is the incredibly abstract or intellectual nature of the "real" world of work and play—the 24/7 world. We often fail to appreciate fully how unreal is the world of floating currencies, speculative financing, Big Data, reality TV, lifestyle advertising, image consulting, and, of course, the ever-present screen. What Baudrillard tells us about this world is what some of us already may know implicitly, namely, that we are living within the closed horizon of technology, or within an idea-world.

There is good reason why many ancient cultures were aniconic. The prohibition against creating images of divine beings and prophets rested on the fear of idolatry. Only those well acquainted with the seductive power of the image understand the propensity to confuse the image and the imaged. We moderns are idol-worshippers. We love our images, so much so that they serve as a template for reordering the world. Our experiences and reference points, our reflections on them, and our self-understanding, are almost entirely conditioned by this reordered reality, by a simulated reality. Seduced by the dynamism of a world organized around the principle of limitless novelty and possibility, we fail to see to what extent this world of infinite potential issues from a very limited and finite perspective on the nature of things. We think we live big and pride ourselves on the openness with which we embrace the future as we progressively enclose ourselves in world of our own making where everything we encounter bears the stamp of our presence.

NOTES

1. A gay science, as I read Nietzsche, is a way of thinking and being in the world that renounces the dreariness of modern rationality and its will to truth, which aims to understand reality as "it really is." What takes its place, as alluded to in the preface to the second edition of *The Gay Science*, is a positive, life-affirming orientation that embraces the embodied human condition.

2. Richard Feynman's playful and idiosyncratic personality is revealed most vividly in his autobiography, *Surely You're Joking Mr. Feynman!: Adventures of a Curious Character*, (New York: W. W. Norton, 1985).

3. The film clip in question can be found on many sites on the Internet. One easily accessible site goes under the title, "What's the Name of a Bird?"

4. Perhaps the most famous example of Plato's employment of myth is his recounting of the Myth of Er in chapter 10 of his *Republic*, 614–21.

5. The nonlinearity of reality is a concept drawn from Ian O. Angell and Dionysis S. Demetis's *Science's First Mistake: Delusions in Pursuit of Theory*, (London: Bloomsbury, 2010). See especially 206, 218–19.

6. Jean Baudrillard, *The Intelligence of Evil: Or the Lucidity Pact*, tr. Chris Turner (London: Bloomsbury, 2013), 13.

7. Jean Baudrillard, *Simulacra and Simulation*, tr. Sheila Faria Glaser (Ann Arbor: University of Michigan Press, 1995), 3.

8. For the curious, the cartoon in question is a Quirit panel.

9. A well-documented recent example of air crew unpreparedness occurred in 2009 aboard an Air France Airbus flight from Rio de Janeiro to Paris. All 228 passengers and crew members died.

10. Molecular modeling is sufficiently established to have dozens of journals given over to the practice. *Journal of Molecular Modeling, Journal of Molecular Graphics and Modelling,* and *Journal of Chemical Information and Modeling* are a representative sampling of the long list.

11. Jean Baudrillard, "The Precession of Simulacra," in *Simulacra and Simulation,* 12.

12. Baudrillard says of Lascaux II, "It is in this way, under the pretext of saving the original, that the caves of Lascaux have been forbidden to visitors and an exact replica constructed 500 meters away, so that everyone can see them (you glance through a peephole at the real grotto and then visit the reconstituted whole). It is possible that the very memory of the original caves will fade in the mind of future generations, but from now on there is no longer any difference: the duplication is sufficient to render both artificial." Jean Baudrillard, "The Precession of Simulacra," in *Simulacra and Simulation,* 9.

13. Baudrillard, *The Intelligence of Evil,* 13.

14. Ibid., 14.

15. Ibid., 14.

16. Ibid., 15.

17. Jean Baudrillard, *The Conspiracy of Art,* ed. Sylvere Lotringer, tr. Ames Hodges (New York: Semiotext(e), 2005), 112.

18. Consider, in this regard, Baudrillard's claim: "We commonly say that the real has disappeared beneath a welter of signs and images, and it is true that there is a violence of the image. But that violence is substantially offset by the violence done to the image: its exploitation for documentary purposes, as testimony or message, its exploitation for moral, political, or promotional ends, or simply for the purposes of information." *The Intelligence of Evil,* 71–72.

19. Jean Baudrillard, *The Transparency of Evil: Essays on Extreme Phenomena,* tr. James Benedict (London: Verso, 1993), 6.

20. Baudrillard, *The Intelligence of Evil,* 150.

21. This quotation is drawn from an interview Jean Baudrillard had with Claude Thibault in *Dialogues,* entitled "Cybersphere: A Discussion with Jean Baudrillard." The interview is available online at: http://www.infopeace.org/vy2k/thibaut.cfm.

22. Nicholas Carr's full length treatment of technology's impact on our cognitive abilities is contained in his *The Shallows: What The Internet Is Doing To Our Brains.* The text is an expansion of his "Is Google Making Us Stupid," an essay first published in the July/August 2008 issue of *The Atlantic.*

Chapter Six

Our Faith

Faith, hope, and things not seen. What do these intangibles have to do with technology? Ours is the age of the tool and exploitable knowledge. No new scientific discovery goes unutilized. If, as happened recently, a new form of light is discovered, follow-up analysis is always action oriented. As lead researcher Paul Eastham said of the discovery above: "But this science is in a very early stage—the next stage is to work out the consequences, how this could be used in everyday life."[1] Of course. We moderns do not adopt the passive stance. Resignation is not our thing. We do not accept what past civilizations assumed was inevitable. As a bearded nineteenth-century revolutionary once put it: "The philosophers have only interpreted the world, in various ways; the point is to change it."[2]

The implied contrast between "interpreting" and "changing" in the above quotation is revealing. The either/or nature of the comparison suggests neither has anything to do with the other, that "thinking" is an altogether different enterprise than "doing." This understanding is as lodged in our collective mind as it is mistaken. As argued here, the real world of technology is premised on a way of interpreting the world. The instrumental character of our age is grounded in an idea, an imagining. Its expression in technologies is traceable to an understanding of the meaning of things that aligns the real with the useful. It is a reading of reality that makes possible the project to change the world.

What motivated this reinterpretation and the emergence of a new worldview? This question was addressed previously in relation to Machiavelli's role in the West's transition to a more activist stance. It was noted that in its rejection of fatalism, the technological world-picture recoiled against a fundamental teaching of the medieval church. But religion factored into the equation in other ways as well. Consider, for example, the fractious religious scene in Europe at the time of the birth of modern science. Religious wars rocked the continent for well over a century in the wake of the Protestant Reformation. Francis Bacon, one of modern science's earliest and most influential champions, was deeply troubled by the threat religious fundamentalism posed to free rational inquiry, especially in the guise of natural philosophy or modern science.[3] An astute political actor, Bacon sought to save science's independent spirit by linking it to a new common project capable of transcending the schisms that were tearing Europe apart in his time. As stated, the project's animating ideal was to relieve the human estate and scientific understanding was to be the means to attaining that end. In Bacon's hands the technological project was conceived for the purposes of expedience. It was to be the new binding agent for a continent racked by a force that once unified it, but no longer—Christian faith. Ironically, this project was intended to serve the same function as the power it sought to sideline, namely, social defragmentation. The technological enterprise was to take on the sociological function of religion, that is, the uniting of peoples behind a common ideal.

If a case can be made that the technological project was conceived to fulfill the failed promise of religion, its end with Bacon also is religiously inspired, albeit with a twist. As stated in his *Novum Organon*, the restoration of science was to play a key part in helping humanity recover from its fallen condition: "For man, by the fall, lost at once his state of innocence and his empire over creation, both of which can be partially recovered even in this life, the first by religion and faith, the second by the arts and sciences."[4] Bacon, it appears, placed on science's shoulders a hefty portion of the burden for creating a "New Earth," a future Eden where humanity regains its lost dominion over creation. Properly understood and deployed, science held for Bacon the power to redeem the world.

But it is a perverse account of redemption we are talking about here. The Bible informs us that it was precisely the desire for wisdom that led to our expulsion from the Garden. All human sin is said to be derivative of the perverse human desire to know. The human estate, in all its hardship, is a byproduct of the fateful decision not to live innocently within God's embrace. But Bacon tells us that knowledge of a sort is a means of returning to something resembling the prelapsarian state that knowledge was responsible for exiling us from initially. Knowledge now is seen as part of the solution to a greater existential problem, not the problem itself. So it remains to this day.

It should come as no surprise to hear that Bacon was an admirer of the ancient philosophical tradition known as Hermeticism or that he likely was an initiate of the esoteric society known as the Rosicrucians.[5] In keeping with the vision of this subterranean movement, Bacon was attracted by the idea of a perfect order, a New Atlantis, and by the prospects of its realization. Modern science was to play a vital role in the making of this new order, if it were to come to pass. We continue to live in a Baconian age to the extent we adhere to a vision of technoscience as a socially cohesive and radically transformative power. With the withering away of the overtly religious component of Bacon's comprehensive plan for the reformation of learning, exploitable scientific understanding has come to represent our one best hope for liberation from the travails of human existence. It is a hope that remains strong.

A 2015 Pew Research Center survey asking American adults their views on science and society revealed that almost 80 percent of the population believes that science has made life easier and holds a positive view of science's impact on the quality of health care, food, and the environment.[6] While not entirely uncritical of certain technoscientific developments, such as genetically modified foods, the numbers reveal that for the vast majority of Americans the conveniences associated with technological advance outweigh the complications that often accompany technoscientific progress.

Prognostications about the future are necessarily hypothetical. Even though past and present studies reveal the public to be sanguine about technoscience's capacity to make for a more secure and convenient life, there is nothing about the contemporary situation that predicts its continuation into the future. Still, barring a catastrophe, it seems likely our collective faith in the power of science and technology to benefit further humankind will persist, despite the fact that we no longer seek merely to assuage life's more vexing challenges. This faith remains remarkably robust after a century of abuse where the principles of rational reordering were placed in the service of the vilest ends imaginable, the Holocaust being the most notable example. But this is not a matter of concern serious enough to have prompted a thoroughgoing reconsideration of the technological project of mastery. The faith is deep.

Some commentators question this faith, or at least its grounds. The French philosopher Luc Ferry is one of them.[7] He tells us that we live amid the rubble of an age where the deconstruction of all previous ideals is total. A rough contemporary, Jean-Francois Lyotard, spoke of the same phenomenon as the collapse of "metanarratives." Society, he said, has lost the capacity to believe in the "big stories" that once supplied it with purpose and direction, a prime example being the emancipation of humankind. The grand ideals of the Enlightenment era are dead, Lyotard concludes.[8]

Ferry agrees, albeit resignedly. In his estimation there no longer are any transcendental ideals capable of serving as the ends of societal development. Projects are no longer possible, as a result, because projects suggest movement toward the realization of exogenous ends and the public imagination has lost whatever capacity it might have once had to sustain belief in such ends. So, from Ferry's perspective, there can be no such thing as the "technological project"—a term employed repeatedly in these pages—because technology is not driven by an end external to its own dynamic. Mastery exists for its own sake, not for the purposes of realizing a higher end. Technological power is its own justification: Its end and its means are one and the same.

Ferry's assessment of the situation is fine as far as it goes. It is true that, if we lived in a time of guiding myths where societal development was capable of being informed by grand narratives, then technology would have been pressed into service to help realize a particular ideal. Yet this is not how technology functions in contemporary society. If technology was a means to the actualization of an end external to its own dynamic, then it would be possible to imagine the conditions that would satisfy the meeting of such an end. But can anyone imagine a situation where these conditions would be met? Is it possible to think of a point at which, having finished the task we humans set out for it, technology effectively would be unemployed? No, we cannot, which only shows that technology is an autonomous phenomenon driven by an inner necessity sanctioned by its creators.

The problem with arguments such as Ferry's is that, according to its logic, we might as well follow Kevin Kelly's advice and do what technology wants because the collective will to direct technology toward ends more respectful of the human good is nonexistent. There is no suggestion here that we are doing technology's bidding because we have "faith" in technology, either. Such faith would presume that technology is perceived as leading to human betterment, a highly contestable assumption in Ferry's opinion. As he says, "Who can seriously believe that we shall have more freedom and be happier because in a few months the weight of our MP3 players will have halved, or their memories doubled?"[9] While I agree with Ferry that no one should fall for the ruse linking efficiency with freedom or happiness, clearly most people see things differently. And this sociological reality carries import. Faith in technology is real. It may be blind or misguided, but this faith is a potent force in the legitimation of technology.

Ferry seems to think that in heeding no voice other than its own technology has devolved into a process without a purpose, making it incapable of serving as a beacon of inspiration and hope. If technology ever had a purpose, so the story goes, it has long lost its way. The visionary quality attending Baconian science has been transformed, with modern technology, to the pedantic business of enhancing efficiencies. In this context, Kelly's heroic effort to mythologize technology can be seen as an effort to supply it with its missing purpose. What explains the perceived need to cosmologize technology if not to compensate for the gnawing doubt that technology is so much sound and fury signifying nothing?

It may be that modern technology appears all but visionless when measured against the sweeping vistas of human possibility opened up during the Enlightenment era, and that the pursuit of greater efficiencies leads to dubious outcomes of the sort Baudrillard outlines. But can we conclude from this that Ferry is right in saying "nobody knows any longer the direction in which the world is moving?"[10] Hardly. Technology has not left us rudderless. It has an end, albeit not the kind of transcendental end that would qualify it as a true end in the eyes of someone like Lyotard or Ferry. The death of transcendental ideals does not spell the death of idealism per se; it only changes its character. Modern technology reflects this change. In the age of technology, development is its own end. The "good" is identified with the efficient and with the conveniences that spin off from the increasingly efficient use of resources. It is a moot point whether the good as understood technologically will lead to human happiness or other higher ends. People are aware technology has a dark side and that technological advance is accompanied with certain drawbacks. They are aware as well that some of these associated problems have proven to be substantial. But just as certain "costs" are associated with doing business, so are the costs tied to technological progress are accepted as part of the overall bargain.

Rationalists like to think that because technology makes no purposive sense it cannot serve to order society or supply it with meaning. But this evaluation presumes our attachment to technology is primarily reason based when it is not. It is not because technology, as argued, is foremost a worldview, and like all worldviews technology is not a rational construct. While argued for in the early stages of its emergence, the securing of the technological paradigm over the centuries now makes redundant the scaffolding once required to advance and legitimate it. Besides, there is nothing about the world itself that dictates that it can be rationally understood only as an exploitable natural resource. Moreover, very few of us, in the midst of the modern paradigm, are consciously aware of our interpreting the world in technological terms. As stated, the power of worldviews lies in their invisibility. History turns into nature with the consolidation of a paradigm shift. "Common sense" informs us the world's energy exists to be harnessed. Our perceiving the world technologically is not a datum of consciousness. Rather, we inherit a background understanding of the nature of things that gives form to our conscious dealings with the world. So worldviews are constructs within which capacities such as reasoning gain their specific shape. What counts as reasoning is therefore ideologically or worldview conditioned. As previously argued, it follows that to the extent reasoning in a technological society is necessarily correlated with purposive or instrumental forms of thinking, the dominant form of reasoning today reinforces the technological. So-called critical thinking is swallowed up by the technological gyre, too. What passes for critical thinking in the present day context remains solutionist thinking: The pressure to think critically is the pressure to find novel ways of solving problems.

Our connection with technology, then, is chiefly affective. Technology feels right to us. We are most comfortable taking our measure of things by the lights provided by technology. Faster, smarter, cheaper, and smaller are adjectives freighted with meaning for us and consonant with our understanding of the real. Our faith is grounded in the perceived truth of the vision of the world revealed to us by technology. It is a perceptual faith, a given set of assumptions about reality that shapes the way we engage the world.

It is reasonable to assume some persons may take issue with the course of the argument presented here. Thinking about technology, they may conclude, does not lead necessarily to an assessment of technology as radical in scope as the one laid out in these pages. For instance, an argument could be advanced that challenges the premise that technology at bottom represents a value orientation. It might be asserted that our technological order is a mature response to the rational interest in species self-preservation, a transhistorical concern. Thus, despite their differences, a tool-using society and a technological society share a common interest in human self-assertion. Hans Blumenberg adopts a variant of this position.[11] There is for him a continuity between the premodern and modern eras, albeit a continuity of problems rather than solutions. This understanding allows one to underscore the linearity of societal development, and downplay the extent to which modernity is the efflorescence of a radical paradigm shift. The apotheosis of this line of reasoning is articulated in Kevin Kelly's *What Technology Wants*, which, as stated, takes our technological society to be just another Hegelian "moment" in the progressive unfolding of a cosmic destiny. I have dismissed this naturalistic reading of technology on the grounds that it offers too easy an apology for the phenomenon it seeks to critically comprehend. Why do we need the Kellys of the world to defend the good cause when we have modern advertising?

There remains, however, another argument in defense of the technological status quo that merits consideration because it represents a more customary challenge to the critical analysis of technology mounted so far. To be sure, one can think about technology and not reach the conclusion arrived at here that the premises underpinning the technological project are misguided, perhaps even fatally flawed. One could, for instance, acknowledge that the efficiency principle which supports in part the technological paradigm is a "value"—and therefore a variable whose existence is neither natural nor universal—but not conclude from this that efficiency and the technological orientation is problematic in the way someone like Baudrillard argues it is.

In fact, an argument can be made that efficiency and associated values serve laudable social ends. Conceivably, one can think critically about technology in a way that not only defends the technological ethos but finds in it cause for celebration. Joseph Heath is one proponent of this way of thinking.[12] Defenses of this sort tend to be sociopolitically grounded, and his is no exception. That is to say, efficiency for Heath is a "good" by virtue of its usefulness within a particular societal context. There is no cosmological dimension to Heath's argument and others like it. There is not a whiff of Kelly in Heath's defense. No hard selling technology by linking it with some kind of universal destiny, for him. Heath's genetic intellectual makeup never would allow him to proclaim, as has Kelly, that a "single thread of self-generation ties the cosmos, the bios, and the technos into one creation."[13] Neither would it lead him to make some grand Baudrillardian claim that the technological worldview runs counter to the cosmic rules of the game. Heath's defense of technology eschews these sorts of grand claims entirely. His approach is decidedly more political than ontological. As he says, efficiency is "the type of value that allows individuals who have fundamentally different goals and aspirations to engage in mutually beneficial co-operation."[14] A Baconian at heart, Heath valorizes technology because it supplies liberal society with a framework capable of producing and sustaining social peace and prosperity. We hold together as a civilization, he argues, because we adhere to a value that transcends those multiple values that divide us. And that value is efficiency.

Efficiency is said to be the highest common value because it plays a pivotal role in helping solve the problem of social cohesion, a problem we noted emerged in the West in response to the religious wars of the sixteenth and seventeenth centuries. Contractarianism was the overtly political response to social fragmentation. If, Heath asserts, the "perfectionist"[15] model of society and its attendant eudaimonistic moral schema was the problem, the solution lay in refiguring the ground rules of social order in a way that accommodated a pluralistic culture. Social contract theory was the answer. With it a mechanism for social order was found in a context that abandoned previous efforts to secure commonality of purpose by means of agreement over the ends of existence. Heath asserts that efficiency is a unifying value that sits well within a pluralistic social framework because it is a means-related, not an ends-related, virtue. Despite variances in persons' life goals, everyone can commit to the overarching collective task of creating a social order that more effectively responds to the varied interests of its actors.

It is not by accident, Slavoj Žižek observes, that politics today is largely biopolitics.[16] With the state's withdrawal from matters pertaining to the securing of the good life, the ethical life, the political "good" was transposed to concerns related to the security and well-being of human lives. And it is within the context of this switch that efficiency shines as a value. For if politics today is biopolitically driven, then politics aligns itself with the task of delivering the goods of peace and prosperity in ways that maximize desired outcomes. In this regard, we can reframe the Cold War as a contest between the world's superpowers regarding the "best" or most efficient means of delivering a modern lifestyle. That the West won the biopolitical war was less a vindication of its values—which were largely interchangeable with those of the Soviet Union—than its preferred means of realizing commonly held values. The upshot is that the West didn't win the Cold War; technology did (or at least its purer realization). The real ideological war was not between competing political ideologies (between liberalism and Communism), but *within* variants of the ideology of technology itself. It was arguably a civil war, a fight whose outcome consolidated the ascendancy of the technological ethic.

For Heath, the story of efficiency leads to the establishment of the most effective institutional means of securing the technological ethos. The liberal welfare state and some version of managed capitalism are these social institutions. Through trial and error, they have proven themselves to be most efficacious in realizing the biopolitical ideal. Finding more effective ways still to align private interests with the public good within the existing political and economic framework is an ongoing task that, Heath asserts, will prove to be "frustrating and unglamorous."[17] But such is the fate of a society that has settled the larger questions of meaning and purpose as we moderns have.

On one level it is difficult to fault Heath's argument for its cautiousness. It is unreasonable to think that efficiency is an inherently problematic value. It is not. What is problematic, however, is the cultural identification of the good with the efficient or the technological alone. Heath appears unwillingly to concede that the outcomes, the very successes of a society given over to the operational ethic may yield results not in our best interests when judged from an extratechnological vantage point. George Saunders's "Jon" is a case study in blowback as it relates to the efficiency principle. He lampoons the view that a social system that "works" necessarily works for us. Baudrillard, as argued, likewise is acutely sensitive to the human cost of technological progress. And it is dire. "Humanity," he opines, "confronted with its own divinized model [of reality], with the realization of its own ideal, collapses."[18] Jon, in the Facility, is emblematic of this collapse. As argued, by extension, we all bear the mark of this declension.

To impress further the dehumanizing or deeroticizing impact of the efficiency ethic let us reimagine Saunders's "Jon" in a different social context. This revisited story we will call "Jen," in honor of its protagonist. Like Jon, Jen spends her time in a facility, only hers is a college campus. It is an attractive place, albeit in the same perfunctory manner most smaller campuses are groomed to appeal. Living in residence, Jen's life is centered around her dorm room, the dorm rooms of friends, the cafeteria, and an on-site shopping complex. Jen has little reason to leave the campus during the term, and rarely does. Like her friends, Jen quickly realized upon arriving on campus that getting a college degree can actually be a lot of fun if approached with the right attitude. The key to success is to remain chill. Knowing she's in it for the long haul, Jen understands the importance of pacing. And having a practical head on her shoulders, Jen quickly came to realize that the best way to manage her college experience is to enroll in as many online courses as she could, which is most of them. "My education is mine to do with what I like," Jen announces to her friends whenever she has a chance, "and I like to learn how and when I like."

Jen's education proceeds apace. Even though she doesn't have to live on campus, given her preference for the virtual classroom, Jen does anyway. It's good for her social life, she realizes. She's a better adjusted person as a result, and that means a lot to Jen. And if she goes through a bad patch or two, as happens when the pressure of exams becomes a bit overwhelming, there's a nearby solution to the problem. Of course, drinking helps, but since you can't drink all the time she's thankful for the REZ-Q dogs. A thirteen-year-old retriever named Bonnie is Jen's favorite—and everyone else's, it seems. So nice that Student Services will bring a furry beast to your dorm room for a minimal charge, too.

One day, however, Jen gets into a funk she can't shake off. Those double shots weren't working. Neither was Bonnie. As best she can recollect, the problem started with a video she watched for an anthropology course she was taking. The forty-five minute film (!) showed a professor talking about a video another professor had made about how important videos were to help learners learn in the modern televisual age. What bothered Jen is that she saw the same film about a film about learning in a history course she wished she hadn't taken. Even though Jen liked the fact that the film was easier to understand the second time around than it was before—a sign that she was learning, she confided to herself—she couldn't help but think she was being cheated somehow. Well, one thing led to another and it began to dawn on her that she wasn't learning much of anything other than how to play the college game. She came to understand all went well as long as she pretended to learn. While this realization was exciting, it also bummed Jen out. She knew one day she'd get a college degree and a job to go with it. But she also knew she was just smart enough to know there was something important she wasn't getting that she should.

Jen struggles over what to do with her unease. After several false steps, however, she takes the plunge and drops out. She understands, more clearly than ever after her departure, that there wasn't much thinking worth its name in an institution nominally given over to the cultivation of thinking. Moreover, with the passing of time, Jen comes to realize that the institutional environment she left behind was actually hostile to thinking's cultivation, that it actively repressed the investigative spirit. Jen's realization hasn't made life easier for her. Years later, she is still dealing with the implications of her understanding. She still finds it all a bit sad at times. But Jen has come to care about learning in a way she never imagined she could. It means something to her now, even though she rarely has as much time to reflect on things of importance to her as she would like.

This Jon-inspired take on the contemporary educational experience puts a finer bead on the problem with technology than Saunders's "Jon," although admittedly with considerably less rhetorical skill. It conveys that Jen's educational experience is to a real education what Lascaux II is to Lascaux I: it is a simulation. It appears to be something it is not. The educational system possesses the trappings of a real education—classes, exams, professors, deadlines, and so forth—while it purges itself of the negatives associated with one. In nominally caring for students, in ensuring the system works for them, colleges forgo the rigor required for a meaningful education. The premium placed on functionality is anathema to the development of education-related competences. That colleges want students to graduate more than they want them to learn reflects the broader cultural milieu and its value preference. Credentialism is key because expertise, or its appearance, is more valued today than thinking.

As stated, a major contention of this study is that thinking about technology is thinking about the force in our lives most responsible for the hijacking of thinking. A technologized educational system is thoughtless, as are other social institutions that succumb to the technological ethos. In showing us why, thinking about technology opens up a path to recovery. How so? To review, we noted that reflective or critical thinking was born with the opening of a space between appearance and reality, between what seems to be and what might be in actuality. This space is tensional. It accounts for what we experience as a desire for understanding, a desire that is premised necessarily on the subsidiary experience of incomplete understanding. So thinking, understood as the love of complete knowledge or wisdom, is knowledge poor but *eros* rich. This much Socrates understood. He experienced the desire to know in a way he could not experience wisdom itself. And, most importantly, he understood what this erotic pull informed him of the nature of the real. There was, then, a knowledge dividend that accrued from his philosophical skepticism. Socrates was a "knower" despite his claims to the contrary. He understood, paradoxically, that the one thing we can know with a measure of certainty is that the truth is withheld from us by dint of our placement in the world we inhabit.

Baudrillard, I argue, makes the same claim. We *think* we know what we are up to with our technological interventions. In an effort to integrate reality, in our commitment to "world-processing,"[19] as Baudrillard felicitously notes, we assume we are fulfilling the hidden promise of the real. As stated, we think with this gesture we are redeeming a broken world. Technology is the answer to the problem of an unredeemed world. It is the truth of the world. A Socratic response to any claim to truth would be to question its premises. What must pertain for the technological project to prevail? What presuppositions regarding the nature of things must hold for technology to be considered a remedy? Are these presuppositions defensible in light of our experience of the world? Questions such as these are rooted in an erotic or zetetic search for meaning and understanding. Their exclusion today from public discourse reinforces the de-eroticized character of our technological society. Things function more effectively when questions related to the propriety of functionality are squelched. A critical function of any ideology, including technology as an ideology, is to banish from the realm of sanctioned discourse lines of reasoning that operate outside accepted parameters of thought. In this sense, technology as ideology is an excuse for the use and abuse of power. To think about technology, then, is to question the belief system that grants technology its legitimacy and authority.

As with Saunders's "Jon," my caricature of a modern university education is intended to bring into relief the cultural drift toward technologism. "College is seldom about learning or thinking anymore," William Deresiewicz bemoans.[20] True, but the interesting question is why. Someone like Deresiewicz fashionably blames neoliberalism or market fundamentalism. In a culture that values monetary values above all else, so the argument goes, an education is reduced in worth to a means of producing wealth. Hence "the purpose of education in a neoliberal age is to produce producers."

This argument is sound on its own terms. But what it does not do is explain what is happening on the educational front in relation to broader societal changes. What, if anything, links thoughtless campuses to driverless cars and Outback Steakhouses? Argued here is that these and other like developments are all of a piece. They are manifestations of the drive to eradicate the evil "other" (or the object) that subverts the technological ideal of pure functionality. In education, the other is ignorance or incomplete understanding. Campuses are hostile to thinking because their stakeholders claim to know thinking's true purpose. The kind of thinking that questions premises, including the premise that conflates thinking with the production of instrumental knowledge, has no place in this realm of intellectual certainty. With vehicular travel, the object is human incompetence. Every driver is a potential bad driver, and bad drivers cause untold suffering in terms of injury, lost lives, and property damage. Drivers therefore are agents of inefficiency that negatively impact the whole of society. Because they are a threat to the biopolitical ideal of human security and well-being, the driver must disappear. Finally, with a dining experience one might find at an Outback Steakhouse, the eradicable object is the place to which the chain restaurant is nominally associated. To eat at an actual restaurant in the Australian outback is an experience few partake in for the simple reason of its geographical remove. It takes time and money to travel great distances. These resources are limited for most of us and therefore are impediments to our enjoying the ambiance of an authentic Outback eatery, whatever that might be. So we fake it. Much like a postsecondary education, we simulate the image of a (nonexistent) reality in a way that makes it accessible to those who do not have the means or inclination to do what it takes to realize the experience of which it pretends to copy.

What does it mean to live in a technological society? It means you ace an exam the contents of which you barely understand, commandeer a vehicle you cannot drive, and dine on USDA prime graded porterhouse in that "Australian inspired" family restaurant down the block. *This* is our reality. This is the world technology has engendered, a world hell-bent on erasing the tension between desire and its satisfaction.

NOTES

1. The quotation is taken from the online CNN article, "Researchers Illuminate the Hidden Properties of Light," available at: http://www.cnn.com/2016/05/17/tech/light-new-properties-discovery/.

2. This often cited declaration in praise of action is the eleventh thesis in Karl Marx's "Theses on Feuerbach."

3. A well-defended analysis of the political motivations behind Francis Bacon's ruminations on science and technology can be found in Kimberly Hurd Hale's *Francis Bacon's "New Atlantis" in the Foundation of Modern Political Thought* (Lanham, MD: Lexington Books, 2013).

4. Francis Bacon, "Novum Organon" II 52, in *The Oxford Francis Bacon* (Oxford: Clarendon Press, 2008), Vol. XI, 89.

5. Good sources on this subject matter include Stephen A. McKnight's *Sacralizing the Secular: The Renaissance Origins of Modernity* (Baton Rouge: Louisiana State University Press, 1989) and Paolo Rossi's *Francis Bacon: From Magic to Science* (Chicago: University of Chicago Press, 1978).

6. This assessment is drawn from the January 29, 2015, Pew Research Center survey, "Public and Scientists' Views on Science and Society," which is available online at: http://www.pewinternet.org/2015/01/29/public-and-scientists-views-on-science-and-society/.

7. Luc Ferry, *A Brief History of Thought: A Philosophical Guide to Living*, tr. Theo Cuffe (New York: Harper Perennial, 2011).

8. As Jean-Francois Lyotard puts it, "Simplifying to the extreme, I define postmodern as incredulity toward metanarratives. . . . The narrative function is losing its . . . great hero, its great dangers, its great voyages, its great goal. It is being dispersed in clouds of narrative language." See *The Postmodern Condition: A Report on Knowledge*, tr. Geoff Bennington and Brian Massumi (Minneapolis: University of Minnesota Press, 1991), xxiv.

9. Ferry, *A Brief History of Thought*, 207.

10. Ibid., 213.

11. For an analysis of Blumenberg's legitimation of modernity, see my *A Discourse on Disenchantment: Reflections on Politics and Technology* (Albany: State University of New York Press, 1993), 68–72.

12. Joseph Heath, *The Efficient Society: Why Canada Is As Close to Utopia As It Gets* (Toronto: Penguin Books, 2001).

13. Kevin Kelly, *What Technology Wants* (New York: Penguin Books, 2010), 356.

14. Heath, *The Efficient Society*, 9.

15. Ibid., 27.

16. Slavoj Žižek, *Violence: Six Sideways Reflections* (New York: Picador, 2008), 40.

17. Heath, *The Efficient Society*, 306.

18. Jean Baudrillard, *The Agony of Power*, tr. Ames Hodges (New York: Ames Hodges, 2010), 82.

19. Ibid., 82.

20. See William Deresiewicz's "The Neoliberal Arts: How College Sold Its Soul to the Market." The essay first appeared in the September 2015 issue of *Harper's Magazine.* It can be accessed online at http://harpers.org/archive/2015/09/the-neoliberal-arts/o.

Chapter Seven

Thinking Past Technology

In the last chapter we explored the nature of our collective commitment to technologism and summarized the consequences of an adherence to the efficiency principle. More generally, it has been argued to date that thinking about technology leads to suspending belief in the technological worldview. It now is time to ask if our deconstruction of technology lends itself to a positive vision of the world to which someone might hold. I already have partially answered this question by suggesting that Jean Baudrillard's dualist account of the real is superior to the reigning paradigm. This alternative vision arguably offers a way of conceiving the world better attuned with our experience of it than the technological world picture. Fleshing out this opposing world-picture is our next task. More remains to be said about ideas related to a participatory understanding of reality, as I have called it.

We can find guidance to this end by referencing the Christian apologist G. K. Chesterton, who once spoke of his faith as responding to a double spiritual need for the unfamiliar and the familiar. He proposed we human beings are best suited to live both with the idea of wonder and welcome, with mystery and comfort, which he identified with a life of "practical romance."[1]

Its religious backing aside, the notion of a practical romance holds an undeniable appeal in the context of this study. That the human condition can be defined in terms of an interplay between opposed yearnings is both insightful and instructive. As argued, the technological worldview is unipolar in that it takes reality be both fully amenable to rational scrutiny and infinitely manipulable. Only "how" questions seem to matter and only because "how" questions seem properly answerable. There is no wonder or romance in the technological worldview apart from the ersatz mystery of not yet knowing what in principle is knowable, which is everything. Clearly, the Richard Feynmans of the world think differently. For them the mystery of not knowing today what one day will be known does not detract from the awe one experiences in the face of the presently not known. This type of response, it seems, sets too low the bar for what qualifies as a mystery. There is a qualitative distinction between a view, aligned with disenchantment, which holds that in principle everything is knowable and one that takes some things as remaining outside human purview. What is missing from the former is the true sense of mystery that attends the realization that reality never fully reveals its essence. What we have instead is a culture that works to make the unfamiliar familiar, the unknown known, the other the self-same. It is all comfort and practicality for us.

A technological culture is by definition romance deficient. There is nothing exotic about "building a better future." It takes work to excite a population about the next step in the Grand March of technological efficiency. The hyping of new inventions such as Google Glass or new iterations of an old technology, such as the iPhone, are testaments to the need to craft desire. Yet we plod along in our collective effort to make real our dreams and ideals. And we are nothing if not industrious. It is for this reason that the most egregious of all transgressions in a technological society is to not increase access to a resource or to let potential remain untapped or underdeveloped. It is not in our cultural DNA to accept that ideals are best left in the realm of the imaginary, that we live suspended in the gap between what is and what could be. It is not part of our collective understanding that efforts to collapse these tensions are inherently problematic. The last century's totalitarian eruptions illustrate most vividly what transpires when we surrender to the political temptation to make real images of perfect order. What we commonly overlook, however, is how this same impulse to whitewash reality is operative today in the purging of the other from the real, which results in the closing of the gaps that otherwise keep reality from being identified with itself. As John Stuart Mill said of the social brand of tyranny, relative to its more overt political twin, it is precisely because the war against the other insinuates itself so deeply into the details our lives that we tend to ignore it or underestimate its impact. [2]

Reinjecting romance into our lives involves a reacquaintance with the object, whose presence has never left us. It entails cultivating attentiveness toward what cannot be assimilated within the realm of "the human." The object, I have argued, lies within us and within the world. It is a property of being as such. It is what lends the world an alien quality, which, among other things, makes a question out of existence. While the world we are born into is our home, it remains in important ways a strange and foreign land. It has been proffered here that technology works to make the unfamiliar familiar under the assumption that the world's alien quality is more illusory than real. We moderns pride ourselves in the belief that the world is, in principle, intellectually transparent and amenable to control. This credo was said to be the cornerstone of our belief system. That the world may forever remain alien terrain is offensive to us, which explains the monomaniacal technological drive to reconcile differences and tensions, to create an integral reality.[3]

The conceit of my argument is that technology simplifies the complexity of lived experience. If we are truly attentive to ourselves and the world around us we understand to what extent the world remains a deeply mysterious place, and how technology papers over this inscrutability by offering us a guide to the universe that, at best, quells our suspicion that we do not know much of what we claim to know. Technology, like any other kind of fundamentalist belief, functions as an anodyne that keeps us from coming to terms with the limits of human knowledge and power. And it keeps hidden the self-understanding that our humanity is tied to the limits of our reach, as much as to the reach itself.

Thinking about technology and its disregard for limits allows us to restore a measure of balance to our lives. It informs us that alienation—the distance that keeps distinct self and other—is not an evil to be exterminated. But incredulity toward the narrative touting salvation through technological advance is just a first step. For it is one thing to see through the hyperbole obscuring the fantastical presuppositions that underpin the technological vision and quite another to know what might lie on the other side of the demythologizing divide. Besides, as important as becoming an apostate of technology may be, like their more straightforwardly religious counterparts, disbelievers of technology remain bound to the notion they discredit. So, if disbelief in the promise of technology is a necessary first step in alienating oneself from a disalienating worldview, it is not the last, at least to the extent one may want to consolidate a position contesting the supremacy of the technological vision.

Seeking ways to engage oneself with the world so as not to reinforce the technological paradigm flows from a re-visioning process that underscores the participatory nature of our placement with the larger order of things, with our immersion in the world. Before explaining further what is meant by participatory, it is important we make clear what "world" means in the current context. This word has been used somewhat casually to this point. At times I have meant by it the world of artifice—the world of our own making—and sometimes the world of nature, the so-called "given" world into which we are born. For present purposes, "world" is to be understood in the widest possible sense as including both the natural realm and the realm of human artifice, or the given world and what we humans have made of the given world.

Observing the worldliness of the human condition means accentuating not merely our inclusion within the world but that this inclusion is constitutive of our very being. We are, in other words, not merely "in" the world but *of* it, as well. At one level, this identification is to be understood literally. Given that the Earth and everything in it are ejecta of supernovae, it should come as no surprise to learn, for instance, that the four most common elements comprising the human body are amongst the six most abundant elements in the universe. But we are of the cosmos in another equally important and fundamental way. This has to do with what might be called our prereflective engagement with the world. That is to say, we are of the world to the degree that what we are as thinking and acting beings cannot be neatly dissociated from the world within which we think and act, including the realm of human artifice. Our condition as embodied beings living in the body of the world means we are fundamentally attuned to our ever-shifting environment. Every time we push we are pushed back. There is no excepting "the human" from the world. Yet we commonly think otherwise. We moderns persist in assuming that with sufficient skill our interventions they will not redound negatively upon us. We are repeatedly shocked and dismayed when they do, as they always do. Nicholas Carr's *The Shallows* is a study that in a more sophisticated cultural environment would not have to be written. It should be obvious that inventions such as Google and the Internet not only help consolidate an established worldview but alter neurological pathways in a manner conducive to its continued hegemony.

A participatory reading of our worldly condition is necessarily holistic. We humans are what we are by dint of our engagement with the world, including the second-order reality we make for ourselves—the realm of human artifice. Accordingly, there is no refuge from the real, no safe haven where we can poke and prod without at the same time being poked and prodded. The reciprocity between (human) being and world means that we are not the sole agents of our perceptions, our actions, or even our thoughts. We are acted upon as much as act and not just in a simple mechanical way, either. It is not as if the world can be likened to a vast container filled with atom-like bits of moving matter whose trajectories are determined by their interactions with other bits of matter in motion. A more appropriate image is that of a "field," as understood by modern day physicists.[4] A field is a continuous medium of energy that extends across time and space and that impacts the behavior of particles in its midst. Particles "feel" fields, physicists like to say, and fields exist to be felt by particles. The Higgs Field, for example, is an invisible energy continuum thought to extend throughout the universe and to be responsible for lending particles their mass. Stated philosophically, the Higgs Field is the *ground* of the material cosmos, the "thing" without which there would be no subsisting things. It is the singular entity from which emerges a multiplicity of distinguishable and observable entities.

Likening reality to a field reinforces the necessary unity of the whole of being. Reality is a plenum, a fullness, from which distinctions are drawn. It is because reality is a plenum that it exceeds any effort to delimit it conceptually. Any "picture" of the real leaves behind a residuum, in other words.

Something always remains behind after any attempt to cut into the fabric of being and secure its essence. This is what the scientific imagination has difficulty grasping. It seeks the final cut, the last articulation that reveals the truth of the real. It keeps digging deeper to isolate the ultimate constituent parts of the physical universe, only to find another, albeit finer image in place of the old. This project retains its legitimacy only if it is assumed that the limits of perception are ultimately surmountable, which in turn presupposes the fundamental transparency of the real.

What accounts for the assumption that we humans possess the capacity to comprehend fully the order of things? How is it that we have convinced ourselves the world can be made to reveal its secrets and submit fully to the power of knowledge? The short answer, proposed here, is that the problem lies ultimately with a misguided worldview. This problem I have suggested is tied to a related misunderstanding of perception. Specifically, it pertains to a neglect of the limits of perception, to its perspectival character, a problem associated with the theme of embodiment. The root of the problem lies with the assumption that perception is a disembodied act, a mind-based phenomenon. The tendency to disassociate perception from its worldly context is understandable. After all, thinking (or cognitive perception) is a withdrawal from the world of appearances, as Hannah Arendt rightly observes. [5] It is an inherently abstractive power and for this reason lends itself to a theory of mind of equal abstraction.

The trap of thinking of the perceiving mind as some kind of generic thinking substance is corrected for by recognizing the body as constituting the nexus linking "self" and world, for it is difficult to conceive of embodied perception in terms other than perspectivally. As positioned, perceiving bodies, the limits of perception are evident in a way they are not when a more abstract representation of perception is in force. Because embodied perception is always a perceiving from somewhere, what is perceived is likewise necessarily determinate. The added benefit of this reorientation is that a new language opens up that enables us to speak of the entire existential context within which the perceiving/perceived event takes place. Perception is no longer a window onto the world but an opening *within* the world or within the field of being that is reality.

The previously mentioned Maurice Merleau-Ponty spoke of this field of being as the world's "flesh." [6] He introduced the concept to describe the worldly condition that must preexist in order for perception to unfold at all. This condition is one of contiguity. Perception, Merleau-Ponty argued, is possible only because the perceiver and the perceived are, at some fundamental level, made of the same stuff. The human body perceives and is perceived by other perceiving bodies, including itself as a perceiving body. Its active powers of perception are inextricably linked to the fact that it itself is a body among other bodies and therefore receptive to being seen, as well.

Like perceives like; it can be no other way. *Flesh* is the term Merleau-Ponty employed to articulate that aspect of reality that positions us in the thick of things.

It is because we are in the midst of reality and not just bystanders that perception is both an active and a reactive phenomenon. It has been noted in this regard that to reach out and touch something is also to be touched by it. As often is the case, the act of touching is solicited by the perceived object. We are invited by the world to press our flesh against its flesh in what amounts to an act of communion. Likewise, although less obviously, to see something is in some primordial sense to be seen by it, to be lured by the thing that ostensibly we visually seek out. To assume otherwise, to think of ourselves as monads who perceptually scout an indifferent and alien world "out there," according to self-generated interests, is a gross simplification of the reality of lived experience and a highly unerotic reading at that. It is an understanding that greatly diminishes the richness, depth, and mystery of our embodied condition.

The participatory picture of reality Merleau-Ponty gives us has been taken up and expanded upon by several key contemporary interpreters. Hubert Dreyfus and Alva Noë are among some of the more prominent. Both focus on the shortcomings of an "intellectualist" picture of reality that prioritizes rational deliberation over other types of skills humans have at their disposal to engage the world. Noë's approach is of special interest here given his forthright critique of modern scientific study of consciousness. Modern science, he asserts, continues to treat consciousness as a strictly brain-related phenomenon and so takes up the challenge of determining how consciousness is generated through neuronal brain activity. Nöe counters with the quip that we are "out of our heads."[7] Matthew B. Crawford, who comes to the same debate from a different angle, speaks of the importance of remaining attuned to "the world beyond your head."[8] In both instances the authors call for an expanded view of what constitutes human consciousness, one in line with basic phenomenological insights regarding embodiment. The upshot is the claim that consciousness is a byproduct of action, of coping and doing. As Noë puts it, consciousness is something we enact through our engagement with the world. This is a contrarian claim, at least from the perspective operative within the mainstream sciences today. It eschews the notion that the "seat" of consciousness is isolable. Parallel to Merleau-Ponty's assertion that perception is a relational activity that conjoins subject and object within the world's flesh, Noë tells us consciousness is an embodied phenomenon that issues from the dynamic interplay between a perceiving subject and its worldly environment. So if it can be said that consciousness is a brain-related function it must equally be said that it is body and world related, too. The world that shows up for consciousness is not an adjunct to awareness but constitutive of it.

The problem with the scientific attitude is that it "manipulates things and gives up living in them." It assumes, Merleau-Ponty wryly observes, that "everything that is and has been was meant only to enter the laboratory."[9] And what enters the lab exits a standardized, abstracted and de-eroticized image of reality, a model of the real whose incompleteness reflects the limits of the analytical framework from which it emerged. To be clear, the problem with technoscience according to this reading is the assumption that it holds a patent on what counts for knowledge, not that scientific thinking does not lead to a form of knowledge. If scientific thinking could coexist in a cultural setting where "living among things" retained its own authority, someone like Merleau-Ponty would have little reason to complain. He has made it abundantly clear in his writings that one can and must cast a critical eye on science, not because science fails to produce its own kind of truth, but because the truth it produces tends to draw us away from a more direct experience of the sensible world. So it is not the artifice of the technological worldview that rankles Merleau-Ponty. We have noted that all worldviews are artificial insofar as they are culturally generated. Rather, it is the "absolute artificialism"[10] of scientific thinking that is unacceptable. Noë would agree with this assessment. Mind science, like modern science in general, conflates "objective" knowledge with thinking from above, which results in the kind of thinking that isolates consciousness within a dedicated organ. Noë correctly asserts there is no empirical justification for this move. His phenomenological positioning is intended to save mind science by having it incorporate into its own self-understanding the centrality of lived experience.

It should be reemphasized before moving on that the picture of reality emerging here is just that, a picture or image of the real. It is an interpretation whose strength lies in its being a reading of reality more fully adequate to lived experience than its technological counterpart. This proposition is not intended to be merely academic. It is raised not simply as a means of dislodging an arguably bad idea—the technological paradigm. The point of thinking about technology is to reflect anew on one's experience of the world. Is it not the case that your eye is as often drawn into the world by virtue of the world's revealing itself to your vision as you self-consciously direct and redirect your gaze? Is there not some kind of interplay between these modes of vision, where what initially "catches your eye" then becomes a focus of visual examination, or the reverse where a self-directed gaze provokes an unexpected stimulus? And what are the broader implications of this "push me, pull me" dynamic? What does it say more generally about the world and our relationship to it?

The problem with the technological paradigm, to repeat, is that it remains firmly rooted in the categorical distinction between the thinking and acting self, on the one hand, and the extended material realm, on the other. While certain recent scientific revelations have forced a reconsideration of the view that the perceiving and acting self can be neatly distinguished from the perceived or acted upon object, this reevaluation has not dislodged the Cartesian mind-set that undergirds mainstream scientific analysis and, more generally, the modern zeitgeist. It has not because if it did the technoscientific project would be undercutting its own authority as the sole arbiter of the truth.

Postmodern theorists have an affinity for quantum mechanics, and for good reason. The attraction lies with the fact that quantum mechanical theory provides them with hard (that is, scientific) evidence supporting the view that the perceiver and the perceived are mutually implicated. Because, the theory goes, the act of measuring reality (at the subatomic level, at least) alters in some fundamental way the thing measured, attempts to draw a clear distinction between "subject" and "object" are unsupportable. In no uncertain terms, then, quantum mechanical theory upends the intellectualist picture of reality that undergirds the technological paradigm. The problem with using quantum mechanics as a guide for reflection has to do with scalability. While we might not be able to pinpoint simultaneously the position and velocity of an electron, no one has difficulty fixing the coordinates of a moving vehicle or any other macroscopic entity. The subject-object distinction appears to reassert itself at scales properly human.

Luckily, there are ways to muddy the Cartesian waters without having to delve into epistemological questions related to quantum mechanics and the like. To this end, I want now to introduce into the discussion of what it means to "live among things" the thought of two individuals who tackle the question regarding the nature of reality in ways that supplement those already provided by Jean Baudrillard and Maurice Merleau-Ponty. Their inclusion into the debate aims to enrich our understanding of what constitutes the quixotic thing called "lived experience" that the technological paradigm works to neutralize.

David Bohm remains relatively unrecognized in his field of expertise despite his considerable intellectual achievements. [11] A physicist by training and a polymath by inclination, Bohm broke the traditional scientific mold by asserting that science is a primarily perceptual enterprise and consequently shares certain features previously addressed in our discussion of natural perception. Informed by revelations within the theory of relativity and quantum mechanics, Bohm came to an appreciation of what has been dubbed the "observer-participant" conception of reality. [12] In contrast with the realist school of thought that underpins classical physics, this antirealist orientation argues against the strict separation of observer and observed. It is a position within the scientific realm that mirrors the participatory outlook espoused by Baudrillard and Merleau-Ponty, and holds interest for us because it provides an argument in support of what it means to live among things.

Bohm regards as unsound the view that science reveals insight into the "true" workings of an independently existing reality and incorporates this contrary stance into his overall understanding of the nature of reality. All of his insights into the nature of reality are heavily inflected with skepticism toward science's capacity to reveal the truth of the real. As with Baudrillard, for Bohm the fullness of reality forever exceeds any given articulation of it. Typically, the suggestion that the world transcends our capacity to capture it conceptually brings with it intimations of hidden realities, immaterial substances, and other metaphysical entities. God lurks in the background. But this religiously informed understanding of transcendence is not the only way to come to terms with the antirealist position. It is possible to speak of transcendence *within* immanence, too. That is to say, reality itself can be seen as structured in a manner that resists comprehension in its totality. The locus of the world's mystery just as easily can be located in the world as beyond it. In the final analysis, it might be impossible to distinguish between the two.

Bohm was especially sensitive to the world's mysterious character. As a scientist, however, he was disinclined to look to some God-like force to account for its elusiveness. His preference was to look to the world itself for an explanation and he found one in a theory of perception. It was by underscoring the perceptual basis of scientific understanding that Bohm broached the subject of nature's transcendence.

Science for Bohm is as much a perceptual enterprise as any other. He argues that, at some level, all thinking is pictorial.[13] Mental imagery is an analog of visual imagery, as perhaps it must be given the integration of our sensory faculties. It follows for Bohm that there is no shortcut to the real itself (or the Real), no world of "pure" or unadulterated ideas outside the cave, no way of apprehending reality other than through its picturing. Because reality, no matter how finely resolved, always takes on the appearance of reality, there can be no finality to the progression of images procured by the scientific imagination. Atoms, neutrons, and quarks have been revealed, in succession, as mere appearances of a deeper reality whose fundamental essence science promises to discover. It will not happen. Today's reality is tomorrow's epiphenomenon, to be sure, but there remains an inbuilt obstacle to the unconcealing of the truth of the real, which has to do with the fact that the preconditions for revealing such truth simply do not exist. The perceptual nature of human cognition, its embodiment within the world it seeks to fathom, means the world never reveals itself transparently, not even to the scientific gaze. While Bohm, like Merleau-Ponty, does not discount the fact that knowledge of a sort is gained as reality is mined scientifically, this understanding, he says, is more a byproduct of the perceptual act than its primary end: "The ability to perceive or think differently is more important than the knowledge gained."[14]

In saying this, Bohm reconceives a scientific understanding of reality in a way that has it dovetail with "living among things." If the essence of reality is predestined to escape us—if reality transcends every effort to seize it, perceptually—then physics never sheds its questioning or speculative core and therefore shares with non-scientific explorations of reality a participatory reach. For if the world does not exist to be possessed, to be made into a manipulandum, what option remains but to live in and with it, to explore ways of articulating the reality of which we are a part?

Bohm permitted himself the luxury of creatively reimagining the nature of reality in a manner that complemented a key scientific insight of his time—namely, quantum mechanics—and better accounted for it. This re-reading posited reality as comprised of a series of worlds within worlds, with the "explicate" order referring to the realm of manifest being, the sensible realm of time and space, and various levels of "implicate" order grounding the "real world."[15] While fascinating, the details of Bohm's understanding of what constitutes reality and how this understanding is tied to an ontological interpretation of quantum mechanics are beyond the scope of this study. So is the issue of the critical reception of his reading of reality. What is important, however, is Bohm's steadfast refusal to identify the map with the territory. He is fully aware of the lure to do just that. Clearly, there is a fit of sorts between the model of reality supplied by science and reality itself, as evidenced by the fact that the knowledge gained of the world through this modeling affords us with a measure of power over the empirical world. The relative congruence between the scientific map and the real comforts us with the knowledge that the world is amenable to human understanding and therefore not entirely alien to human interests. In this sense, technoscience satisfies one half of the dual spiritual need Chesterton identified as practical romance.

The problem is that it does so at the expense of the other half of the equation. By remaking reality in our image the "romance" component of the spiritual dynamic withers. It remains underdeveloped because our picture of reality is taken to be one with the territory it maps. Science is not perceived as just one way amongst others of coping with reality but as the path to revealing the truth of the real. Consequently, it takes on the aura of a holy crusade, as does the putative task of relieving the human estate, technology's proper end.

As argued, the technological project aims to humanize reality. We ease the burdens of existence by reworking the world in ways that make it less alien to our purposes. But the comfort taken in humanizing reality comes at a cost. The price we pay is seeing our imprint in everything we encounter. Otherness disappears. Hannah Arendt put it nicely when she said, referencing Werner Heisenberg, "From this he concluded that the modern search for 'true reality' behind mere appearances, which has brought about the world we live in . . . has led into a situation in the sciences themselves in which man has lost the very objectivity of the natural world, so that man in his hunt for 'objective reality' suddenly discovered that he always 'confronts himself alone.'"[16]

The fact remains, however, that no human endeavor to account for reality can be wholly adequate to the task. Thinking otherwise leads to the conundrum cited above, where the effort to get a hold of reality results in our losing it. The only way to stay real, then, is by resisting the temptation to conflate interpretation and text, to come to terms with the illusion of the real, as Baudrillard puts it. It is a paradoxical proposition if there ever was one and not lost on Bohm who incorporated this insight into his own intellectual efforts. Knowing the systemic limitations of any reading of the real allowed him to think both creatively and playfully. It kept him open to the irresolvable mystery that undergirds the real.

Political philosopher Eric Voegelin is similarly concerned with questions regarding the being of reality and our placement within the larger scheme of things.[17] And he also sees the human condition in terms of a dual spiritual need for the familiar and the unfamiliar. Voegelin articulates this double need within the frame of experience called human consciousness. Like a good phenomenologist, he notes that consciousness is always and necessarily an awareness *of* something. Because consciousness by definition is intentional, the expression "pure consciousness" is nonsensical.[18] For this reason, consciousness is always concrete. Voegelin means by this that consciousness is a fundamentally embodied phenomenon. Reality opens us to us as beings with bodies and, therefore, within specific spatiotemporal and cultural contexts.

What interests Voegelin about our conscious experience of reality is what this experience reveals about our self-understanding and the world that shows up for us. But Voegelin is no phenomenologist of the Merleau-Pontyan variety. Voegelin deviates from the latter regarding what constitutes the experience of reality. Reality for Voegelin is not primarily the world of direct sensual experience. It is neither simply a "thing" (the sensible world) nor a conceptual representation of the sensible realm—an "idea." Rather, reality as Voegelin understands it is an experiential truth, as we may say of time. Because time is something we live, it accords with our understanding of the true and the real. Yet it cannot be pointed to or intellectualized in a way that does justice to the experiencing of it. Likewise, reality for Voegelin is not something that can be satisfactorily accounted for by giving an inventory of existent things. Nor is reality simply the idea of the whole of reality. Rather, reality is the lived experience of being suspended between the immanent world and a world beyond. Simply put, for Voegelin, reality shows up as something that points beyond itself. Reality contains within itself intimations of transcendence. It is an open field of being.

What precisely this something points to or where it resides is not of ultimate concern to Voegelin. Often, and understandably, the experience of transcendence is codified as the divine, God, or one of its cognates. But for Voegelin the label or symbol attached to the experience is less important than the experience itself, the sense of order experienced in the face of a luminous reality and the ways this order have been articulated across time. Human beings throughout history have experienced this openness toward transcendence or toward "the divine ground of being," as Voegelin sometimes phrases it. [19] And they have done so with various degrees of symbolic sophistication. Despite the variances in these articulations, however, what unites them is the sense that consciousness is a participatory act and that reality constitutes an order that elicits an erotic encounter.

Voegelin is mindful that experiencing reality in its fullness can be corrupted, and often is. These deformations occur when openness toward transcendence is cut off. When this happens, the image of wholeness or completeness linked to the ground of being is repositioned as a realizable good. In recent history, these deformations have taken the form of radical ideologies. Voegelin has outlined extensively how visions of earthly perfection (i.e., the Third Reich, or post-political communist societies) have been mobilized to transform reality to accord with these images and always at the cost of great human suffering. [20] I have argued, in partnership, that the technological world-picture shares with radical ideologies the same goal of creating something akin to heaven on earth, with associated costs just as dire.

Voegelin's "old world" erudition, coupled with his mystical intellectual leanings, lead many to pigeonhole his thought as an eruption of an antimodern, religious crank. That is a mistake. As previously maintained, the notion of transcendence need not be interpreted along narrowly theological lines. Baudrillard, we have seen, speaks to the transcendent character of reality without invoking traditional religious imagery. The same can be said of Camus. Merleau-Ponty, similarly, articulates a way of addressing the issue of transcendence without broaching the subject of metaphysical faith. What these purveyors of the participatory paradigm touch upon, variously, is the "moreness" of reality. Consciousness, for them, and for us, is more than a registering of and reflecting on external stimuli. What is understood by the term "reality" evokes more than what appears simply and directly to the senses. This moreness, this surfeit of being, if we are attentive to it, is what lends the world its distinctive character. It accounts at once for the perceived realness of the real and our sense of living real lives.

Herein lies the balance that is a practical romance. The world is more mysterious than we are given to believe, a curiousness that can be appreciated only if we acquaint ourselves with the virtue of living among things, as opposed to living from without. Ironically, as observed, living among things requires accepting the otherness of reality—its objectness, its transcendence, its duality. It demands that we remain mindful of the extent to which the world is not ours simply to do with what we like, but a domain within which we live erotically charged existences. For all its existential might, the technological vision remains cramped because it shields from view our suspension in the thick of things—even the thick of technological things. In our mania for mastery, for instance, we lose sight of the delicious irony involved in our capture by our tools of domination. Baudrillard delights in revealing that seduction always gets the last laugh. In denying the seduction of the real, we mobilize forces to operationalize reality, only to have this recreated reality seduce its creators into ceding its authority to the mechanisms of management. This is why, in a doubly ironic move, Baudrillard contends the only way to regain a measure of humanity in a technological society—to live among things within a world that pushes things around—is by refusing to be lured by the efficiency principle. "Now you must fight against everything that wants to help you," he says emphatically.[21]

To be human for Baudrillard is to resist being seduced by the "Empire of Good" and its cadre of professionals who ostensibly care for human well-being.[22] This is sage advice. Jon, we have seen, regains his personhood by rejecting the infantilizing efforts of his caregivers. Likewise, registrants in today's educational system who wish to be students of true learning must learn first to spurn the blandishments of the custodians of learning, and the simulacrum of an education to which their siren calls lead. Across the board, Baudrillard asks us to disengage the power structures that work to collapse the tension between desire and its fulfillment. Technology's deeroticizing impulse is antihuman. It follows, oddly enough, that conserving our humanity requires retaining an appreciation of the object or the other than human—the inhuman. This appreciation for otherness extends even to the otherness in us. As argued here, to be human is to experience being suspended between desire and fulfillment, potential and actualization. It is our fate, in short, to be incomplete beings. The shadow of a human life, the negative without which there can be no positive, no accomplishment, is the life unrealized or the life missed out.[23] It is the limits of a human life, self-imposed or otherwise, that provide the ground for its reach and meaning.

Baudrillard relates a story that, as he puts it, "says it all" regarding technology's penchant for disregarding limits and its consequences. It involves a man walking in the rain with an umbrella tucked under his arm, who responds, when asked why he chooses not to open it: "'I don't like to feel I've called on all my resources.'"[24] Arguably, the man with the unopened umbrella is *eros* personified. As with the Greek god *Eros*, born of Porus (resourcefulness) and Penia (poverty), the figure in Baudrillard's story is self-divided insofar as he refuses to utilize an available resource. In this regard he is the embodiment of both fullness and lack. One senses the man with the umbrella is quite capable of employing it, and sometimes does, but not in every opportune circumstance. He therefore inhabits a space between a capacity to act and the necessity to act. He is free. The spirit of technology, in contrast, is slavishly bound to its master. A technological society is compelled on every front to realize potential, to liberate an inchoate nature by extending indefinitely its instrumental hold on reality. Such a society's operating system cannot countenance the conditions under which *eros* flourishes.

Living among things is anathema to the technological ethos. The spirit of technology wants nothing to do with the thing-like character of the real. As Albert Camus put it, "We live in the time of great cities. The world has been deliberately cut off from what gives it permanence: nature, the sea, hills, evening meditations."[25] What can a stupid and silent nature tell us? Nothing. There is no sustenance to be found, no guidance to be gained, from incorporating into our self-understanding the cosmic backdrop against which we forge our existence. For us it is all history and no nature, all forward movement and little appreciation of nature's indifference to our plans for it. There is a scene in the original Indiana Jones film that serves as a wonderful symbolic broadside to the blather of today's technoidolaters. In it a sword-wielding tribesman confronts Indiana in a public square. The attacker brandishes his weapon with a flourish that suggests Indiana's imminent demise. Indiana momentarily considers his predicament, then, almost dismissively, draws his gun, kills the assailant, and continues with his adventure. The dark humor in the scene is the same humor one finds in witnessing any well-laid design gone to ruin. Ultimately, reality remains indifferent to human designs. It is this indifference that ensures "the world always conquers history in the end," Camus observes.[26]

Sadly, but not unexpectedly, the insight gleaned from an understanding of the limits of human intervention rarely is reinjected into the realm of historical action in a way that might inform it. Any potential balance between action and understanding is undermined in a cultural context that conflates understanding with know-how.

One of the many ironies that attend our neglect of nature and the contemplative pose, and our wholesale embrace of history and the activist stance, is the disappearance of history's actor (i.e., the self) and therefore history itself. In neglecting the inhuman, and engaging oneself in the historical work of humanizing nature, history arguably ends with the liquidation of the humanizer. Differently put, by relinquishing our capacity to live among things, by turning the world and ourselves into manipulanda, we create a world that no longer needs us. The economy of technology renders "the human" redundant.

In light of this consideration, the conceit of George Saunders's "Jon," its imaginative limitation, becomes apparent. The presumption embedded in the text is that humans like Jon are required to discern consumer "likes," when there is no good reason to assume they are. If the Facility represents a step up the efficiency ladder, relative to today's consumer feedback practices, the next logical step is the elimination of its "weak link"—humans. Facility 2.0, when it arrives, will house integrated circuitry, not people. So Jon, even if he felt entirely at home in his bubble, was wise to leave the Facility with Carolyn. There was never a future for them there, anyway. Algorithms, analytics, and Big Data can read popular taste more effectively than humans, especially when existing standards of "human" taste are always already conditioned by agencies of control.

In parallel, we may ask: Why learn? Why do we need humans to think when we have "better" means to affect the same outcome? How long will it be before we come to grips with the redundancy of an educated mind, and hence of education itself? Or have we already, with the proliferation of the contemporary learning-free educational institution, and businesses such as No Need to Study, which encourage students to "imagine all your online courses being taken by an expert for you."[27]

The entire trend to render ineffective human agency, whether it be human skillfulness relative to robotic performance, or human intelligence relative to artificial intelligence, must be seen for what it is. Thinking about technology helps us gain a long view of this trajectory, with the added bonus of providing a means of its assessment. To date I have provided two opposing trains of thought with regard to technology and technological development. On the one hand we have Kevin Kelly, whose analysis is representative of those who think about technology within the bounds of technological thinking. I have argued this type of thinking amounts to not thinking about technology at all. What Kelly and his ilk excel at is *selling* technology. He exhorts us to buy in. In this regard, Kelly's defense of technology is merely an excuse for the use and abuse of power. The title of his latest book, *The Inevitable*, is proof enough of his motive.[28]

Against this reading we have presented another view that not only remains skeptical of technology's promise to deliver on all things good, but renounces the entire technological enterprise on the grounds that delivering on all things good is precisely what renders technology deeply problematic. For Baudrillard, who represents this opposing view, technology poses a mortal threat to humanity. This challenge, importantly, has nothing to do with the concerns of many who are anxious about our technological future, such as climate change, nuclear devastation, and the like. The real danger for Baudrillard and for us is spiritual, even metaphysical.

In an important sense the age of technology is also an age of spiritual crisis. Precipitated by technological advance, a crossroads has been reached that is forcing us to decide whether or not we need the world. Baudrillard, we have seen, has rightly pointed out that the "best of all possible worlds no longer needs us."[29] What technology wants is a posthuman world. And, for the reasons given, it appears technology is getting its way. Humans are adapting to their marginalization with remarkable ease given the momentousness of the shift. The question is whether or not this transition to a posthuman future remains uncontested. Arguably, the moment has come where we are called upon to respond to technology's calling. That technology doesn't need us is a given. The really important question is this: Do we need the world? Is "living among things" an experiential reality of sufficient existential pull to counter the allures of a simulatory universe?

NOTES

1. As G. K. Chesterton explains, "But nearly all people I have ever met . . . would agree to the general proposition that we need this life of practical romance; the combination of something that is strange with something that is secure. We need so to view the world as to combine an idea of wonder and an idea of welcome. We need to be happy in this wonderland without once being merely comfortable." *Orthodoxy* (New York: Dodd, Mead & Co., 1908), 4.

2. As John Stuart Mill puts it in the "Introductory" to his *On Liberty*, "Society can and does execute its own mandates; and if it issues wrong mandates instead of right, or any mandates at all in things with which it ought not to meddle, it practises a social tyranny more formidable than many kinds of political oppression, since, though not usually upheld by such extreme penalties, it leaves fewer means of escape, penetrating much more deeply into the details of life, and enslaving the soul itself." *On Liberty* (Indianapolis: Hackett Publishing, 1978), 4.

3. This reference to "integral reality" hearkens back to our discussion of Baudrillard's understanding of technology in chapter 4.

4. The best summation of a field I have come across reads as follows: "A new concept appears in physics, the most important invention since Newton's time: the field. It needed great scientific imagination to realize that it is not the charges nor the particles but the field in the space between the charges and the particles which is essential for the description of physical phenomena." Albert Einstein and Leopold Infield, *The Evolution of Physics: From Early Concepts to Relativity and Quanta* (New York: Simon and Schuster, 1938), 244.

5. Hannah Arendt says the following: "The primacy of appearance for all living creatures . . . is of great relevance to the topic we are going to deal with—those mental activities by which we distinguish ourselves from other animal species. For although there are great differences among these activities, they all have in common a *withdrawal* [author's emphasis] from the world as it appears and a bending back toward the self." Hannah Arendt, "Volume One: Thinking," *The Life of the Mind* (London: Secker & Warburg, 1978), 22.

6. The most sustained treatment of the concept of "flesh" is contained in Maurice Merleau-Ponty's *The Visible and the Invisible* ed. Claude Lefort, tr. Alphonso Lingis (Evanston, IL: Northwestern University Press, 1968).

7. The expression "out of our heads" happens to be the title of Alva Nöe's phenomenological critique of consciousness studies. See his *Out of Our Heads: Why You Are Not Your Brain, and Other Lessons from the Biology of Consciousness* (New York: Hill and Wang, 2010).

8. See Matthew B. Crawford's *The World Beyond Your Head: On Becoming an Individual in an Age of Distraction* (Toronto: Penguin, 2015).

9. Maurice Merleau-Ponty's, "Eye and Mind," in *The Primacy of Perception*, ed. James M. Edie (Evanston, IL: Northwestern University Press, 1964), 160.

10. Ibid., 160.

11. David Bohm's relative obscurity might have something to do with his storied past. In the immediate post-war period, David Bohm was an assistant professor at Princeton University and was working closely with Albert Einstein. Then, in 1949, Bohm was called before the House Un-American Activities Committee where he pleaded the Fifth and subsequently was arrested. Losing his position at Princeton, Bohm relocated first to Brazil, then to Israel, and finally to the United Kingdom where he settled at the University of London's Birkbeck College. Bohm sums up his intellectual outlook nicely in the following passage: "I would say that in my scientific and philosophical work, my main concern has been with understanding the nature of reality in general and of consciousness in particular as a coherent whole, which is never static or complete but which is an unending process of movement and unfold ment." David Bohm, *Wholeness and the Implicate Order* (UK: Routledge and Kegan Paul, 1980), introduction, x.

12. The apotheosis of observer-participant theory is John Wheeler's notion, "genesis by observership." See "Does the Universe Exist if We're Not Looking?" in *Discover*, June, 2002. Available online at: http://discovermagazine.com/2002/jun/featuniverse.

13. As David Bohm says, "Here, one has to emphasize that the act of reason is essentially a kind of perception through the mind, similar in certain ways to artistic perception, and not merely the associative repetition of reasons that are already known. Thus, one may be puzzled by a wide range of factors, things that do not fit together, until suddenly there is a flash of understanding, and therefore one sees how all these factors are related as aspects of one totality (e.g., consider Newton's insight into universal gravitation)." David Bohm, *Wholeness and the Implicate Order* (London: Routledge Classics, 2002), 17.

14. This quotation was accessed from a John Horgan article on Bohm entitled, "Last Words of a Quantum Heretic." *New Scientist*, Issue 1862, 27 February, 1993, 42.

15. For an overview of the two orders, see Bohm's *Wholeness and the Implicate Order*, 186–90.

16. Hannah Arendt, "The Conquest of Space and the Stature of Man," in *Between Past and Future: Eight Exercises in Political Thought* (New York: The Viking Press, 1961), 277.

17. In Voegelin's words, "The ground of [human] existence is an experienced reality of a transcendent nature towards which one lives in tension. So, the experience of the tension towards transcendent being is the experiential basis for all analysis of such matters." For Voegelin, then, we humans experience reality as opening toward the whole of existence, an experience that transcends purely mundane perceptual phenomena. See *Conversations With Eric Voegelin*, ed. R. Eric O'Connor (Montreal: Thomas More Institute, 1980), 8.

18. Voegelin's position on consciousness is revealed in the claim: "An attempt to withdraw consciousness from its ontic context, to squash the world and its history and to reconstruct it out of the subjectivity of the I, is not a matter of course." See his "On the Theory of Consciousness," in *Anamnesis*, tr. Gerhart Niemeyer (Notre Dame: University of Notre Dame Press, 1978), 34.

19. Voegelin's references to the transcendent or divine ground of being are many. An especially well-focused analysis, however, can be found in "The Consciousness of the Ground," in *Anamnesis*.

20. In the regard to the political deformations endemic to modernity, see Eric Voegelin's *The New Science of Politics* (Chicago: The University of Chicago Press, 1952), especially chapter IV, "Gnosticism—The Nature of Modernity."

21. Jean Baudrillard, *The Agony of Power*, tr. Ames Hodges (New York: Semiotext(e)), 88.

22. Ibid., 88.

23. See, in this regard, Adam Phillips' *Missing Out: In Praise of the Life Unlived* (New York: Farrar, Straus and Giroux, 2012).

24. Jean Baudrillard, *The Illusion of the End*, tr. Chris Turner (Stanford, CA: Stanford University Press, 1994), 101.

25. Albert Camus, "Helen's Exile," in *Lyrical and Critical Essays*, tr. Ellen Conroy Kennedy (New York: Vintage Books), 150.

26. Albert Camus, "The Wind at Djemila," in *Lyrical and Critical Essays*, 79.

27. See *The Atlantic* essay (November 4, 2015), "Cheating in Online Classes Is Now Big Business." The article is available at: http://www.theatlantic.com/education/archive/2015/11/cheating-through-online-courses/413770/. The cited quotation is lifted from the *No Need to Study* website, at: https://www.noneedtostudy.com/myclass/.

28. Kevin Kelly, *The Inevitable: Understanding the 12 Technological Forces That Will Shape Our Future* (New York: Viking), 2016.

29. Baudrillard, *The Agony of Power*, 81.

Chapter Eight

Living Among Things

Do we need the world? Does living among things matter anymore? Thinking about technology helps raise questions of this sort. It also helps answer them. Our complicity in the technological project suggests we do not. The flip side of the fact that technology does not need us is that we do not need what technology effaces—the object. Living among things, in contrast, means living in the midst of an obdurate reality. It means living in the world as it is. Baudrillard, for one, stresses that living in the world as it is is not how we moderns navigate reality. Through our technological interventions, he contends that "we attempt to control not that which is, but that which, in the name of this assumption, ought not to exist."[1] We have seen that for Baudrillard ridding the world of evil is a fool's game. Wishing evil or the object not exist and working toward its eradication does nothing to resolve the fundamental duality that underpins the real. So despite our efforts to transcend reality, to move beyond the world and the rules of the game, we remain mired in what it is we seek to escape. Technology or no technology, reality prevails.

There is a huge difference, however, between seeking control over what is and over what ought not to exist. With the latter or technological route, we move toward the establishment of the unified state of things called integral reality. A chief "accident" of the enterprise to integrate reality is the collapse of the human. Posthumanism is the price paid for violating the real. Seeking control over what is constitutes an altogether different tactic. The aim here is not dominate the object-world, but to live within it in a way that affects outcomes deemed desirable to the establishment of a life worth living. This alternate tactic aims lower than the technological approach in that it accepts the built-in limits of efforts to systematize or humanize reality, but in doing so preserves the integrity of the relationship between human and world.

In the previous chapter we stated that thinking past technology requires an understanding of what it means to live among things. The articulation of this understanding was laid out primarily in theoretical terms. Here we will examine this same theme in a more pragmatic context. What does the admonition to live among things mean in terms of actual life experience? Especially, how might it be possible to live an engaged existence within a cultural milieu that actively suppresses our erotic attachment to things? These are important questions for many reasons, but perhaps most of all for the reason that cognizance of the abstractive power of technology—its capacity to distance us from our embodied condition and sever our erotic attachment to reality—leads invariably to calls to "get real" and to an explanation of the form this grounding may take in the midst of a technological order.

The irony is that "getting real" today amounts to getting unreal, to reacquainting ourselves with the illusion of the real. How can this reacquaintance be affected in our everyday comportment? What behaviors lend themselves to living among things? In general terms, the answer lies in an actional orientation that might be called "painterly," as opposed to linear. By painterly is meant an approach to living that finds ways to remain open to an encounter with things in their givenness. The point is not live life as if one were coloring between the lines of a preestablished order. Rather, the ambiguity that attends the world as it presents itself ought to be met with behaviors that take advantage of the slippage between human and other. It follows that there can be no prescriptive set of rules, no program, that flows from this exercise in thinking about technology that detail specifics with regard to action. But we are not left in the lurch as a result.

Some headway can be made outlining the contours of the good life as determined by insights gleaned from our thinking about technology. The good news is that no one is asking anyone to leave entirely the world they know, the world of comfort, which in this day and age we identify with modern technology. Obviously, technology is here to stay. Thinking about technology will not make it go away. What this thinking has the capacity to do, however, is alter perceptions regarding its value as a common project. Ideally, from the vantage point of this study, reevaluating the value of technology leads to loss of faith in technology's guiding ethos. This is not an inconsiderable achievement, for while living today is inconceivable outside a technological context, this existential datum does not carry with it the imperative to define ourselves within the parameters of the technological. What is on the table is not an either/or proposition. It is absurd to think, in the contemporary context, that one can be either "for" or "against" technology. It is axiomatic that all of us want a welcoming world, welcoming to us humans, that is, and this means a world reordered to better suit our needs. The technological apparatus serves this end. The question of balance remains, however. Does our desire and need to reconfigure reality admit of any restraint? Is it comfort all the way down? Once we start on the path toward humanizing reality are we necessarily doomed to inhabit a worldless world of the sort Kevin Kelly envisions? It certainly looks that way. However, in the final analysis, whether or not we hit the target is less important than our drift toward it. The Singularity might not be in our future, but what it represents remains the ideal our technological order strives to realize.

We fail to acknowledge the developmental arc of technological progress not because we have reflected on it and rejected the conclusion that technology has a built-in telos or endpoint, but because we have not thought about it at all. The unthinkingness that attends faith can be redressed only by questioning faith, which in our context means probing exactly what it is we are up to when we embrace the latest time or labor-saving device. It is arguable that, if invited to think about the unthought, the charm of the entire enterprise begins to fade. We realize the merchants of welcome and comfort are over-selling their wares and that we would do well to rebuff their invitations.

It is not comfort per se that needs revisiting. The attraction of the idea of comfort is incontestable. What is less certain, however, is the appeal of the idea of wonder, and how this idea (should its attraction hold) can be acted upon in the context of a technological culture. As argued, our capacity to admit wonder into our lives is hindered by the constitution of the world we have built for ourselves. Who wants to take the long route to acquire a competence, a route that makes demands on one's time and patience, when resources are limited and shortcuts abound in the form of user-friendly technologies?

The temptation is great to resist paths of resistance, and often for perfectly understandable reasons. Chief among them has less to do with helping us cope with the practical affair of managing our busy lives than with our living in a social environment that forces upon us a reliance on technology. It goes without saying that it is next to impossible to extricate oneself from the technological matrix, even if one wishes to. That is because we have inherited a social system that by definition is not, strictly speaking, our own. The general contours of the modern world have been set for by historical processes going back at least to the Enlightenment. What freedom we have as individual actors is constrained by the structures of action imposed on us by the guiding values that inhere in our received social order, the primary value being freedom, especially negative freedom or freedom from constraint. The good life, for us, is a life where things go our way, where we realize our desires with a minimum of resistance. Technology's operative ethic, as argued, is the efficiency principle, which is aligned with developing and refining the means to create delivery systems that facilitate the realization of perceived needs. This is the attitudinal template we inherit as moderns. It is a given. The question is this: What to do with it? Further embrace the user-friendly ethic or give it a shake?

There is no third option. Either we blithely follow the virtual yellow brick road and meet our technological fate, however it turns out, or we take the skeptical path and suspend belief in the belief that technology can save us, and move unsteadily ahead. In keeping with the worldview of the ancient Skeptics, the position advanced here is one that forefronts a searching, questioning attitude. Not disbelievers in the truth, Skeptics were circumspect only of our capacity as humans to possess definitive knowledge. The skeptical attitude, directed toward technology, suspends belief in its animating ethos or truth claim, which holds that reality can be rearranged in ways that permit the excision of evil. That is well and good, but as argued we must go further than merely withhold faith in technology's saving power. Simple disbelief is nihilistic and for this reason capricious. Principled disbelief is something else entirely. Principled disbelief in technology rests on an understanding that the technological worldview misrepresents the world its images and acts upon. To be effective, any serious questioning of technology must be guided by an alternative vision of what constitutes the real. The alternative picture presented here is of an erotically charged world, a vision that carries with it certain guidelines for action.

So, what, to return to our original question, are the practical consequences of disbelieving technology? If the realm of human artifice reflects how the world is perceived—if reconstructed worlds and worldviews are mutually implicated—then what kind of actional world emerges from the participatory worldview offered here as an alternative to the technological world-picture? The answer, in short, is a way of being or acting in the world that reclaims the tension between subject and object, between the actor and the acted upon.

Several disclaimers need airing before speaking to this chapter's main concern. First, everything offered here as insight into ways of reenergizing our erotic connection with the world has been said before in other contexts. My only hope in retreading ground is that whatever understanding may be forthcoming resonates with readers more than it otherwise would given the preceding analysis of technology. Second, the suggestions below are not presented systematically. There is no internal logic to their ordering. Neither are they siloed conceptually. Rather, the recommendations are casually presented and, to various degrees, mutually reinforcing. Finally, it needs be stated that no attempt has been made to provide an exhaustive list of sanctioned advice. My only intention is to adumbrate the actional orientation of someone who embraces what technology rejects—a world with limits.

Don't Just Do Something, Stand There!

This first piece of encouragement might seem to belie a claim to practicality. Getting real and doing nothing appear to be mutually exclusive demands. Of course, they are in the context of a technological society, which is why doing nothing is the kind of doing that needs more doing today. But the do-nothing-ism advocated here is not to be confused within inaction per se. The modern world is action oriented in the specific sense of valuing purposive forms of action. To *do* something is to work toward accomplishing a tangible goal or advancement. The identification of doing with goal-oriented behaviors disin-centivizes the kind of leisurely activities that serve no prescribed end. There is much to be said for idle thought, engaging in a casual conversation with a friend or stranger, reading another chapter of an absorbing novel, or taking a leisurely stroll. Actions of this sort might not accomplish any overt goal-driven task, but that is not to say they are not meaningful actions. Just the opposite. These divertissements are the stuff of life. It is precisely when you are not "on task" that you are likely to be less in your head and more open to the world around you.

Take the Crooked Path.

If you look at a foot-worn path in a park or field, you will notice it never carves a perfectly straight line. Unconsciously, we humans deviate from the geometrical. The shortest path between any two points might be a straight line but the shortest route doesn't seem to jibe with our natural proclivities. There is wisdom in this. It has been stated repeatedly that technology is about efficiency. Technological thinking is geometrical thinking: It seeks the straightest, most efficacious, means of realizing ends. The problem is that the mean-ends technocratic mentality, which favors "the one best way," stands at odds with the human need for diversity and spontaneity.

Old cities, even the remnants of old cities in the core of newer ones, generally are deemed more attractive than more modern conurbations be-cause they are less obviously purpose built. Likewise, one might argue the most appealing lives are those that weave in the pursuance of their goals. An interesting and adventurous life is a life lived in the absence of rigidly preset ends and means. Such a life is responsive to transgressions—to challenges and roadblocks on a chosen path—and what, after all, is an adventure but a transgression one doesn't regret?

Show Me Your Hands.

The spirit of technology is spiritual. We can blame René Descartes for setting the tone for modernity and, by extension, the age of technology. The modern existential mood is exemplified by the search for certain knowledge. Des-cartes's search led him to conclude that the only thing we can know with certainty is that whose existence cannot be doubted, and the only thing that cannot be doubted, he concluded, is the act of doubting itself. The "thing" that does the doubting, for Descartes, is the thinking thing called mind—an immaterial "substance" set against the realm of material substance. So, the reasoning goes, the only real and certain thing is the immaterial thing called mind. Mind is focal.

We continue to live in the shadow of Descartes. As noted, technology is mind obsessed: we moderns likewise are mind-maniacs. We privilege the brain and its technological surrogates as the singular source of thought, and the source of all things good and worthy. We believe smart technologies and Big Data will make for a better world. In privileging mind, we said, the object is necessarily overlooked. There is a failure to account for what works at cross-purposes to the mind's designs.

Part of what it means to "get real" is to find ways to reengage the object, which means, as well, finding ways to reinvigorate the subject. If an over-reliance on the controlling mind goes a long way to explain why we lead such abstract, ethereal lives today, then reclaiming the real necessarily entails repossessing our embodied condition. For it is as beings with bodies that we most conspicuously confront the perversity of things. Tapping a screen to download an app is not an embodied act, at least not of the sort intended here. To the contrary, our haptic interactions with our devices lead us away from understanding the real as resistant. These interactions disengage the body from an active engagement with the real. They make us redundant as corporeal beings.

Having the world (or its facsimile) at your touch may hold certain functional advantages, but what it teaches us about our relationship to the greater order of things is less than salutary. It misleads us into overestimating the powers of human agency, of mind over matter. The best way to set aright this unreasonable presumption is to engage the world more concretely—flesh to flesh, so to speak. Working with one's hands in the performance of a craft or practice is the most obvious route to this end. As Martin Heidegger rightly noted, we think with our hands as much as with our minds. More than this, he adds, "[O]nly a being who can speak, that is, think, can have hands and can be handy in achieving works of handicraft."[2] So thinking, according to this reading, is intimately tied to working with one's hands. Thinking, perhaps at its best, is a coping that is engaged most productively by a kind of doing.

Matthew Crawford, of *Shop Class as Soulcraft* fame, is someone who has devoted his intellectual career to reclaiming the real along these lines. As he puts it, in a more recent text, "to pursue the fantasy of escaping heteronomy [otherness] through abstraction is to give up on skill, and therefore to give up on the possibility of real agency."[3] The sentiment expressed in this passage is in tune with Baudrillard's and my own. The fantastical technological vision, by disregarding the object, undermines the subject as well. The possibility of rekindling real human agency therefore hinges on our engaging the object. Getting real means getting dirty. It means committing yourself to practices that take effort to become proficient in and which by their nature prevent one from mastering absolutely.

It is optimal, then, to devote energies to developing the skills that facilitate becoming adept in whatever practice or practices one may be drawn to. Learn to play the guitar, or continue honing your musical talent. Pick up a paintbrush. Take up gardening or a sport. All that matters is engaging a resistant reality in a way that encourages one to develop the skills of worldly interaction.

If you need any more convincing, consider the following: higher-ups in some of the world's leading technology giants, such as eBay, Google, Apple, Yahoo, and Hewlett-Packard, send their children to schools than ban high-tech gadgetry. No computers, no screens of any kind are found in Silicon Valley's Waldorf School of the Peninsula, to name but one.[4] What do these tech savvy leaders know that we do not? And if what they know is good for children, then why is it not also good for adults? How did we ever come to think otherwise?

Nix the Expertocracy.

The disappearance of the subject means in large part the vanishing of the subjective or first person perspective. I introduced Eric Voegelin into the technology debate to underscore the centrality of "lived experience" (or living among things) to the human condition. Of course, it is impossible in the final analysis not to see the world through your own subjective experience of it. The point, however, remains that this experience is always a filtered experience and that not all filtered experiences are the same in terms of what gets laid over the original ground of experience, which Voegelin equates with an appreciation of the mystery of being, or, in the language preferred here, the otherness of things. Increasingly today, what intercedes between you and your perception of reality is the expert, who almost always is an expert of the technical sort. Whether credentialed experts focus their attention on monetary, dietary, parental, or myriad other issues, their intention is directed typically toward informing you how to be a more effective or efficient actor. There is a double loss here. One, as a consumer of expert-based information, an ethos is reinforced that conflates action solely with purposive action. Understanding is identified with the "how to" mentality associated with the technological worldview. Acting "well" within this parameter means acting in ways that maximize the efficiency with which one expends available resources with sharp boundaries demarcating the actor from the acted upon. Lost in the process is the sense of openness to that which is that thinkers such as Voegelin argue constitutes the core of the human experience. Two, the closed world of human experience is mirrored and secured by an equally closed world of social relations that has us become clients of a managerial class of efficient experts. These experts are as much a technology as any hand-held device, and relying on them for life advice is no different than depending on a GPS system to navigate around town.

Advice that encourages sidelining the expertocracy need not be seen as growing out of a libertarian bias toward rugged individualism. Aristotle was right to say that anyone who has no need for other human beings must be either a beast or a god.[5] Humans by nature are social creatures who rely on others for a whole host of reasons. Relying on others for guidance in making life decisions is not in question. Neither is the value of scientific understanding per se. Rather, what is at stake here are the consequences of ceding to the cognoscenti (or, even worse, to celebrity culture) suggestions for shaping and comprehending one's own experiential reality. Over-relying on the expertocracy steals from you the validity of your own experience of the world. Even if informed, this dependence plugs you into the circuitry of popular opinion. Worse yet, it reduces you to a node in a communication system that has you adopt a generic point of view based on general or abstract claims concerning how best to function. All of this tends to rob persons of the courage it takes to trust (as prudence dictates) what their own encounters with the world inform them about the world.

In an age obsessed with integration and synergies, with closing the spaces that separate things and processes, it is important to do your bit not to be a bit player in the systemization of the world. It is impossible to find out if you know more than you think you know unless give it a try, in which case, even if you lose, you win.

Alienation—It's a Good Thing.

Alienation is an idea that gets no respect. Nothing is more antithetical to the technological mindset than the cluster of connotations associated with the term alienation. Estrangement, divorce, separation, and disconnection are never nice words, but especially for us moderns, for whom dealienation is our very raison d'etre. In suggesting that alienation is a good thing I am not saying we ought to work at creating divisions where divisions never existed, or deepening schisms that already exist. Rather, I am suggesting that you consider our bias toward everything that is implied in the expression "conflict resolution" as indicating a commitment to an unreal and unrealizable ideology of rapprochement. If alienation is a good thing, it is only because alienation lies in the heart of the real. There is, as a result, nothing moral about the claim regarding alienation's goodness. It simply is. The problem is that the totalitarian attempt to integrate or unify—the technological temptation in its essence—disregards the nature of what it tries to unify, reality itself. As stated, it is simply an existential fact that the world is constituted in a way that precludes its radical systemization. Including us, as part of the overarching order of things. So it is, to repeat, that the world does not offer its meaning to us unambiguously, and that bad things happen to good people and good things to bad. The disjunction between what we want out of life and what is on offer for is something we are stuck with. We either come to terms with this "reality" or we do not.

Living among things, then, entails reconciling oneself with the alienated human condition, or at least working toward doing so. There is nothing here that says one must relish this state of alienation, even if this were possible, only that one is better served acknowledging it and accepting what is. A false sense of what lies within the realm of the possible leads invariably to disappointment (and later, cynicism) when reality intercedes, as it invariably does. Wisdom lies in accepting that the disappointments, frustrations, and anxieties of life are not merely ineradicable, but constitutive of the human experience. Technology's spirit informs us that we can have our cake and eat it too, that we attain our full humanity by resolving the tensions in our lives that prevent the real world from matching the world of our dreams. That is nonsense. With every step we collectively take toward achieving an earthly paradise, we move closer to the prelapsarian condition of Adam and Eve—to re-animalized beings.[6]

Wisdom resides in resisting the forces in your life that would have you believe happiness is within your reach. It is precisely because we don't get out of life what we want that lends it existential heft and meaning. "Heaven is a place, a place where nothing, nothing ever happens," David Byrne once warbled.[7] Exactly. So why reach for nothingness?

Eros, Eros, Everywhere . . .

We moderns live in an erotically charged world whose nature we fail to acknowledge and fold into our own self-understanding. Rectifying this oversight is key to the business of getting real and living post-technologically.

Eros is the yang to alienation's yin. Plato understood well that we humans are by nature desiring beings and that our desiring nature is premised on our not having what we yearn to have. In the *Symposium* he has Diotima explain this condition mythologically by citing that *Eros*'s character (and ours, by extension) is forged in the zone between nothingness and fullness.[8] As a result, we are driven by an existential lack toward fulfillment. This yearning takes many forms. The sex drive is one obvious example. But so is the desire for any kind of power that distracts us from our needy human condition. Material wealth, political power, military fame, athletic accomplishment, all can be seen as manifestations of the erotic drive to quell the unrest at the center of our being.

Technology, too, can be understood as an erotically driven enterprise. There is no purely rational reason that accounts for the modern will to knowledge. Understanding the world scientifically is not a necessary precondition for human existence. The rational superstructure that is modern science and technology rests on a base that is not, strictly speaking, rational. Scientists are driven by impulse. They are erotic beings at bottom, whose devotion to science is a particularized manifestation of a force we all partake of as human beings.

Why is this important to know or otherwise reaffirm? Because it puts into perspective the technological will to power and the world the will to mastery creates. It was argued at the outset that to understand technology requires framing an analysis of technology that situates it within a broader canvas. It was said we need an outsider's view of technology. The introduction of erotics helps provide this more encompassing interpretive framework. What needs affirming is that we are playthings of forces much larger than ourselves and, therefore, that we are destined to remain seduced by the world. The great game in which we participate sees to it that our need for completion or totality will never be satisfied. It is because our technological culture is grounded in an ethic that promises what cannot be delivered that our efforts to realize to "perfect" reality are doomed to backfire. Any civilization obtuse enough to work toward realizing absolute power is working simultaneously to eventuate its own disappearance.

The lesson that ought to be learned from an understanding of erotics, as applicable to individuals as to entire civilizations, is to play the game that presents itself to us. Strive toward excellence. Especially, given the advice offered above, strive to engage skillfully in hands-on "worldly" activities. If you engage yourself properly, you will be taught through the nature of these experiences the humility that comes with a devotion to craftsmanship. The world will let you know, despite the talents you acquire along the way, who or what is in ultimate control.

Get Geopoetical

Thinking is not unrelated to doing. How we perceive things is correlated with how we interact with the world and reorder it. Geopoetical thinking is thinking directed toward the object, or, specifically, toward the object cum geosphere. Scottish thinker and poet Kenneth White describes it as "an attempt to read the lines of the world."[9] White's geopoetics is aligned with Baudrillard's revelation that the object remains "the horizon of my thinking."[10] For both, the most demanding and rewarding of all experiences is being attentive to the given world of appearances. Reading the lines of the world brings one face to face with the inexhaustible depth of the text called the phenomenal world. It involves trying to see past the layers of accumulated meaning that have accreted onto the world and prevent us from seeing it in its raw particularity. Like artists who attempt to see the world anew, with the eyes of a child, it is important that we work toward opening our eyes to the majesty of the sheer givenness of things.

Is this asking too much? Only if it were not already the case that all of us, at moments, engage in geopoetical reflection. There are times in most persons' lives where circumstances conspire to elicit a repose that has them confront the ineffable thereness of the world. Often, scenes of natural beauty trigger these meditative moments. Now, it may be that only a few can give voice to this type of experiential reality. Not everyone can be a Walt Whitman or a Friedrich Holderlin. But everyone has the capacity at least to experience the lines of the world, a not inconsiderable achievement.

My advice here is to be mindful of one's powers of geopoetic reflection and, to the extent possible, help cultivate these powers. Work with this latent capacity to deepen your sense of awe toward the world. Seek out those moments where the world outside your head slaps you on the side of your head. There is no better therapy for the choice-weary than to experience being chosen by the world to indulge in its presence.

Attention: Attention Deficit and Hyperactive Disorder

At minimum, almost everyone today suffers from a technology-induced borderline case of ADHD. No one I know who speaks honestly and who is old enough to have witnessed the digital revolution has not experienced its attention-related effects. Plugging into a hyperreal virtual world has both real benefits and costs, but it is these costs that should concern us most since they tend to undermine whatever benefits might accrue from living in the information age.

Plato understood that democracy, as a regime type, neither was capable of the greatest good nor the greatest evil because it tended to produce democratic souls that haven't the resolve to accomplish anything of greatness. The Internet has democratized the modern soul more than Plato ever could have imagined. We tweet. We chat. This goes viral, followed by that. Today's passion is tomorrow's . . . what? Who can remember?

A distracted culture dissipates its energies: It loses focus. And without a capacity for focused attention, we become flotsam and jetsam in the technological tide. Moreover, in the process we lose the acumen needed to understand why an engaged mind is a prerequisite for the kind of attentive life advocated for here.

Focusing on why focusing is important will not work for those already suffering from an attention deficit. The best way therefore to extricate yourself from the attention deficit trap is to partake in what Albert Borgmann has called "focal things and practices."[11] Similar to what Hubert Dreyfus and Sean Dorrance Kelly have argued in their *All Things Shining*,[12] focal experiences are those moments of ordinary life where one is cast into the kind of existential mood that elicits sustained attention. Focal practices are liberating in their power to draw us out of ourselves and into the world in an engaged manner. Dinner table conversation, beachcombing, participating in or watching a sporting event, any number of quotidian experiences are capable of delivering persons from their scatteredness.

Again, cognizance of the problem is a precondition for its remedy. In an age that where the "erotics of attention" is given short shrift, one must work at engaging in those types of practices that facilitate focused attention.[13] The object is good for you. There is something deeply therapeutic about the getting out of your head by means of sustained attention on the other. Knowing this is an aid to cultivating your life in ways that allow for these practices to take shape.

Remember: I-It

We noted that by humanizing reality through our technological interventions we are removing the possibility of encountering the real, or meeting the world on its own terms. In the phraseology of the Jewish philosopher Martin Buber, the technological disposition tends toward the promotion of "I-It" relationships, where the world shows up for us as something to be used for human purposes.[14] I-It relationships therefore are monologic, not dialogic. This understanding is in keeping with the notion presented here that our utilitarian culture is a product of a particular (human) inclination to see it in these terms. The technological worldview ignores the other, we said, and so prevents us from establishing a more meaningful encounter with world around us.

It has been argued that a serious critique of technology must question the identification of the real with its image. But in one respect it may be most prudent to retain or even deepen the technological stance. And that is toward technology itself.[15] Keeping our relationship to our technological gadgetry thing-like is central to the demythologizing of technology. It is important that we relate to technology, especially our digital devices, in ways that keep at the fore its status as a tool. It is important that we adopt a strategy that has us *use* technology, that we resist the temptation to lend to technology a dialogical power. Keeping distinct in our minds exteriorizing kinds of relationships from those that involve an engagement with the other (i.e., other persons or the world as other) is critical in an age that works to flatten this distinction.

To conclude, I wish to make clear that the guidelines listed above are akin to what Thomas Hobbes once called "dictates of reason."[16] That is to say, they encapsulate the kinds of attitudes and behaviors that follow reasonably for those who no longer live under the spell of technology. In this regard, they are less precepts than a sketch of the types of worldly engagement one would expect of a technoskeptic.

NOTES

1. Jean Baudrillard, *Passwords*, tr. Chris Turner (London: Verso, 2003), 86.
2. Martin Heidegger, *What is Called Thinking?*, tr. J. Glenn Gray (New York: Harper Colophon Books, 1968), 16.
3. Matthew B. Crawford addresses the theme of de-skilling as it relates to a technological culture in his two major works, *Shop Class as Soulcraft* and *The World Beyond Your Head*. The quotation cited here is taken from the latter work. See Matthew B. Crawford, *The World Beyond Your Head: On Becoming an Individual in an Age of Distraction* (Toronto: Penguin, 2015), 253.
4. The trend toward hi-tech free schools in Silicon Valley was first reported in the *New York Times* article, "A Silicon Valley School That Doesn't Compute," published in 2011. It is available at: http://www.nytimes.com/2011/10/23/technology/at-waldorf-school-in-silicon-valley-technology-can-wait.html?_r=0.
5. Aristotle, *Politics*, 1253a.
6. My reference to "reanimalized" beings is meant to evoke Alexandre Kojève's character-ization of life at the end of history, where "the Subject" is no longer "*opposed* [author's emphasis] to the Object." Kojève's argument is very much in line with the one made here (with the aid of Baudrillard) that technological progress leads to human regress. See Alexandre Kojève's *Introduction to the Reading of Hegel*, "Note to the Second Edition."
7. The lyrics are taken from the song, "Heaven," which appeared on 1979 Talking Heads album, *Fear of Music*.
8. Plato, *Symposium*, 201d–204c.
9. The phrase "an attempt to read the lines of the world" is taken from the *Scottish Centre for Geopoetics* website, at: http://www.geopoetics.org.uk/welcome/what-is-geopoetics/. An excellent starting point for anyone wants to delve into Kenneth White's theorizing on geopoeti-cism is *The Wanderer and his Charts* (Edinburgh: Polygon, 2004).
10. In one of his last writings, Jean Baudrillard encapsulates the key ideas of his thought, and situates "the object" at the forefront. Because, he says in *Passwords*, he "wanted to break with the problematic of the subject," Baudrillard turned his attention to the object, although, as noted, rethinking the object has repercussions for conceptualizing the subject. Baudrillard, *Passwords*, 3.
11. Albert Borgmann addresses the theme of focality in several of his works, but perhaps best in *Technology and the Character of Contemporary Life: A Philosophical Inquiry* (Chicago: The University of Chicago Press, 1984).
12. Hubert Dreyfus and Sean Dorrance Kelly, *All Things Shining: Reading the Western Classics to Find Meaning in a Secular Age* (New York: Free Press, 2011).
13. The phrase "the erotics of attention" is Matthew Crawford's. See chapter 10, "The Erotics of Attention," in *The World Beyond Your Head*.
14. As Martin Buber explains: "The man who experiences has no part in the world. For it is 'in him' and not between him and the world that the experience arises. The world has no part in the experience. It permits itself to be experienced, but has no concern in the matter. For it does nothing to the experience, and the experience does nothing to it. As experience, the world belongs to the primary word *I-It*. The primary word *I-Thou* establishes the world of relations." (italics author's) Martin Buber, *I and Thou*, tr. Ronald Gregor Smith (Edinburgh: T. & T. Clarke, 1947), 5–6.
15. Clearly, there are instances where adopting an I-It relationship with technology is nei-ther appropriate nor called for. For instance, no accomplished musician can afford to relate to his or her instrument as a mere tool. This being said, the general claim being made here stands. Combatting our continued integration into an integrating reality requires a mental form of pushback, a form of self-awareness that brings to the fore the technical apparatus.

16. At the end of chapter 15 of his *The Leviathan*, entitled "Of Other Lawes of Nature," Hobbes speaks of his previously cited "laws" as "dictates of reason" to clarify the fact they are but "conclusions" of already established premises regarding the conditions conducive to keeping the peace. Hobbes underscores these nominal "laws" are not commandments but reasoned deductions of preestablished principles. In the same vein, the suggestions offered in this chapter follow naturally, as it were, from the adoption of a post-technological position.

Chapter Nine

Less Is More

Perspective matters. What you see depends on where you stand. This is equally true of visual perception as seeing with the mind's eye. I have tried in this study to present a big picture view of technology, an alleged under-visualized picture. This expansive intellectual image required we step back from the immediacy of the technological order. The goal was to look at the phenomenon called technology as if we were not part of the thing under examination. The purpose of this exercise was to loosen the hold on our imagination of ingrained perceptions regarding what technology is and does. The ultimate goal was to see technology afresh and to gain a better sense of the true import of our civilizational commitment to the technological ideal.

The significance of perspective to this study can be illustrated by recounting a personal story. Years ago an academic colleague was on a globetrotting adventure. A few months in, I received a postcard (in the days before email) from him informing me that Hegel was right—history had in fact ended.[1] The planet, he intimated, had fallen under the spell of technology. While there were early adopters, laggards, and even a few holdouts, the writing was on the wall: the Chinese are as beholden to the value of efficiency as Americans, and Shanghai and New York City are just variants of each other. The homogenizing effect is ascendant and it is simply a matter of time before the entire globe is remade in the image of the West.

From the high altitude vantage point of Hegelian thought a more than strong case can be made that the planet indeed has fallen under the sway of a monolithic vision of the real. Taking a wide-angle view, the differences that distinguish a Singapore from a Rome all but disappear. A case can be made the similarities between the forces that shape the material lives of most of the world's population outshine whatever political and cultural differences separate them. To fetishize these differences, as many do today, is to ignore the fact that these differences are differences without a distinction, so the argument goes. Whatever variances exist are merely variations on a unifying theme, as vanilla, strawberry, chocolate are to ice cream. Doubtless, critical thinking about technology requires the kind of imaginative positioning that facilitates the broad view. It serves the purpose of allowing for the articulation of the basic lineaments that give shape and form to what is being examined, which, for us, is the topography of the technological order.

If the sole aim of this discourse on technology were simply to apprise readers of the nature of the technological order, this big picture perspective likely would suffice. But my intention was never to stop at convincing readers why it is important to think about technology in a way that might have you reconsider its legitimacy. It has been periodically reinforced that thinking about technology also is driven by a conception of the good life that the technological ethos arguably undermines. So contrariness, per se, has never been a motivating factor in my critical assessment of technology. As stated, deprecating technology for its own sake is nihilistic. The aim of the endeavor has been to think about technology in order to reclaim what it means to think more generally, and to have this thinking address ways of coping with the technological order that ameliorate its most pernicious effects. How can you live among things in an age of mastery over things? How can you live a life of welcome and wonder in a time where resources almost entirely are given over to the task humanizing the real? These are the kinds of questions that prompted our venturing into the technological thicket and must be kept at the forefront of our inquiry.

And so we descend from the Hegelian heights and survey the same terrain from a lower altitude. When we do, we see there are considerable and meaningful variations in the technological texture of various cultures that cannot rightfully be ignored. To illustrate, I again will resort to personal experience, this one postdating the one cited above by several years. The low altitude vantage point from which my observations will be drawn is, in this instance, literal. But first the backstory. Over a decade ago I had the opportunity to travel through India for a month and, like many before me, I found the country in equal parts fascinating and frustrating. With regard to the latter, an afternoon spent at a bus station in a Rajastani desert town stands out most vividly. I remember thinking, as I shuffled along amid the heat, the dust, and the noise, whether the entire outfit had sprung from the ground the night before, so chaotic was the scene. Efficiency, I understood then, was not a universal value. The locals, it appeared, were less failing at the task of running an effective operation than not even trying. Needless to say, I felt very Western. Weeks later, as our aircraft lifted off the runway in Mumbai, the organized chaos that is India made its final impression as the shapeless mass of squatter camps edging the airport slowly faded from view.

Approaching North American soil at the other end of the journey, the image that materialized out the aircraft window could not have been more different. Below loomed a motherboard writ large. It did not take much imagining to envision the complex of freeways, apartment buildings, and storage facilities in view as so many diodes, transistors, and microprocessors—as integrated circuitry. And the ride home did nothing to offset the initial impression. I imagined myself an electron, one of numberless others, being shunted from one pole to another.

It was all very antiseptic. It was all "good," as Baudrillard might put it, and it was unnerving for that reason. Anyone who has spent time in a technologically impoverished culture understands in their bones that too much of a good thing is not a good thing. Why do people who help modernize regions within the developing (read: technologizing) world so often become enchanted with the people they meet and places they come to know? Why does the attractiveness of a place often appear to be inversely related to its level of technical sophistication? Why are so many eager to return? Because they experience, vicariously, the wisdom that inheres in a way of life not given over entirely to the ethos of technology. They sense that there is comfort to be found in meeting the world on its own terms. That working to create a welcoming world does not require the wholesale remaking of the real but a reconstructing that stays within the boundary of the real.

Big picture critical analyses of technology that remain wedded solely to the high-altitude perspective tend to encourage a kind of fatalism that further entrenches the ethos they reject. Heidegger, for all we know, might be right in saying only a god can save us from our technological fate,[2] but claims regarding the futility of resisting the technological juggernaut are unnecessarily defeatist. One doesn't think about technology either to win or lose a war against it. One thinks instead to effect understanding, and changes in understanding are never without practical import. We must remember that thinking is a kind of doing and that thinking technology anew opens up approaches that run counter to the technological flow. As noted, these modes of action denote ways of navigating the world that bypass what today is like as the "natural" course of things. To conclude, if technology rightfully can be called a dematerializing or abstracting power, itself grounded in a big-picture view of reality that holds everything in principle can be known, then resisting the technological tug means reclaiming the centrality of specific, nonreproducible life experiences. Getting real means getting down, in both the literal and metaphorical senses of the word.

If our faith in technology remains unflinching—if we do not find ways of resisting the technological thrust—then we will continue with our descent toward a merger with our machines, our algorithms, and our integrated control systems. Invariably, a posthuman future lies ahead. This fate is not something we moderns are predisposed to reflect upon. Arguably, this aversion can be traced to a naive trust in an essentialist human nature. The belief is that technological advance augments human capacities in a way that leaves untouched their basic character. The argument goes something like this: as "natural" beings, we acquire through our interaction with the world the ideas and skills needed to become more effective actors. Technology amplifies our natural coping skills; it adds to our baseline skill set. The upshot is that technology renders the human superhuman.

This line of reasoning assumes quantitative change is unrelated to changes in quality. More is simply and always more of the same. But nature is rife with processes where incremental changes in a property of a substance eventually result in a phase change that alters fundamentally the substance's structure or character. Baudrillard draws on this understanding in his discussions of evil and reversibility. It is simply not true that having greater access to more information necessarily results in more informed actors, as it typically expected. To the contrary, a case can be made that data glut changes fundamentally the character of the information gatherer. There is an argument to be made that, as in an inflationary economy, a surfeit of information lowers its value. Too much of anything reduces its perceived worth and increases our indifference toward it. This law of value applies to information, to the image, to a university education, or to any other "commodity" circulating within a social system.

The "more is better" ethic has us believe the smarter technologies become the better served we will be. We are told that smart technologies make for smarter people, as in more effective or efficient actors. With their aid we will gain access to more choice while saving more money, time, and other valuable assets. Again, what is presupposed in all of this is that these smart technologies merely enhance some natural and abiding capacity we have to maneuver about the world. They are believed to supplement a basic substrate of human competency. If we were to question more resolutely this assumption, then we would be more inclined to ask ourselves what costs might be incurred for their adoption.

The current misdirection of attention is evidence of a kind of blindness on our part. As already argued, we assume our acting into nature has no consequences for how we are acted upon by nature and our artifactual world.[3] We fail to acknowledge the participatory character of consciousness and what our partaking of the reality we experience says about the relationship between perceiver and perceived. What it tells us, and what we fail to appreciate, is that we are as much here for the world as the world is here for us. Our being and the world's being are mutually implicated. This means, to repeat, that we cannot lay claim to knowing reality objectively, even in principle. The participatory character of consciousness seals reality's fate as an illusion, a fiction. Which is to say that to the extent we "know" reality, we know it symbolically or refracted through images of the real. All knowledge, then remains at bottom poetic or allusive. Scientific understanding is no exception. It has no special status as the arbiter of truth. Rather, science is just another way of coping with the world. This is what we have forgotten in the age of technology. We assume reality is real and we attribute to modern science the power to reveal the real. But it is precisely this understanding that is in error. We live under the illusion that we are the disillusioned ones, the first to have discovered reality. We think we have resolved the historical struggle to "get reality right." The mythopoetic age is behind us.

This presumption needs to be challenged. Arguably, all we moderns have managed to accomplish by jumping on the technological bandwagon is substitute one illusion for another. The problem with technology, we said, is not that the world cannot be perceived in technological terms but that we assume this perception amounts to more than a reading of the real. We fail to understand that reality is an illusion. This is our dilemma and redressing it requires keeping in play the illusory quality of the real. The poetic cast of mind displayed by Camus, Merleau-Ponty, and Baudrillard can be understood only in light of their rejecting the governing intellectualist paradigm. The real for them is not rational, at least not fully. The language of science is not complete speech. Reality's operationalization through technology does not exhaust its being. There remains throughout the irreducible mystery of things. So life is more confounding and profound than as interpreted by the technological imagination. The same can be said of consciousness, which is arguably the ultimate known miracle. But it is a wonder that goes largely unappreciated because consciousness, like everything else today, is not "real" until it is rationalized and subjected to scientific scrutiny. The experiential reality called consciousness is less captivating to us than unraveling the "problem" of consciousness.

Voegelin, we have seen, is a bedfellow of Baudrillard's, although admittedly of a strange sort. Being a participatory phenomenon, consciousness for him can experience the world from the inside alone, through the formulation of symbols of meaning. Reality in an objectivist sense is an illusion for Voegelin, too, and like Baudrillard, disaster ensues if we try to realize—literally, to make real—the order of reality as we conceive it. The real for both Baudrillard and Voegelin is ineffable. It is simply a mistake to assume access can be gained to any ultimate reality. Yet it is an error we not only commit repeatedly but compound by acting on. For the principals of this study, the attempt to realize the real, to close the gap separating the symbol from the symbolized, distorts lived experience in an unacceptable manner. Technology (itself an ideology) and modern ideological thought are the leading forces of distortion today, for Baudrillard and Voegelin, respectively. Bohm, too, we have seen, was perturbed by a parallel tendency within the sciences proper to close off lines of rational inquiry that did not adhere to the reductionist strictures of "legitimate" investigative study. He understood that no form of inquiry, including scientific reasoning, is exempt from the participatory character of conscious thought.

As important a corrective as the alternative to the technological world picture may be, it remains an alternative largely without a public profile for the simple reason that we live in a technological universe. Kelly is right to call our social order the technium. We have replaced an originary nature with a second nature, an artificial ecosystem nearly as comprehensive as the original it supplants. Technology is the medium through which we operate and by means of which we gain our bearings. To the extent technology constitutes our "world," it shapes perceptions in a way that underwrites and reinforces the truth of this world, the object of our belief. There are comparatively few challenges to this claim to truth in contemporary society. And those challenges, when mounted, are typically labeled reactionary, idealist, fatalist, or related pejoratives. This is not to say there are not deep-seated anxieties about the technological order but that these concerns rarely issue in a penetrating analysis of the technological phenomenon. What remains hidden from view is the thinness of the system's underlying vision of the real.

It has been said that the order imposed upon reality always reflects the understanding of the real that dominates at a given time. Our technological landscape is no exception. Today's "on demand" zeitgeist mirrors not only the belief that the world exists to serve our collective needs as a species but our personal wants and preferences as well. We demand the world be reconfigured in a way that has it obey our command. As a result, the given world splinters into a multiplicity of reconstituted worlds responsive to the interests and desires of its manifold "users." Apps are but the latest technological refinement in personalizing our capacity to have the world do our bidding. There will be more. In all of this we are being conditioned to experience "reality" on our terms, not its. The objectness of the real dissolves in the process and along with a taste for resistant reality. Nothing could be more antithetical to the participatory ethic than living in a way that suggests the world is there for us and not us for it. Nothing could be more opposed to the participatory orientation than the belief that humanity is bettered by finding progressively more effective means of having reality yield to human control, to have it effectively disappear and we along with it.

This exercise in thinking about technology opened with a recounting of a discussion between Socrates and a friend on the subject of love and beauty. The topic of conversation, a *human* dialogue, was echoed in their response to its natural setting. The bare-foot friends delighted in their surroundings, as revealed in Socrates's comment: "And what a lovely stream under the plane tree, and how cool to the feet!"[4] Thinking about love and beauty for them was never an exercise in mere cogitation. The inner world of thought was never divorced from the experiential realities of everyday life. Love and beauty were real for them. They spoke about it only to give voice to their experiencing the world as an arena of seduction. In being moved by the world, they lent it expression. And it is because words and ideas *mattered* to people such as Socrates that their words still resonate with us some 2,500 years later.

The dialogue between Socrates and Phaedrus did not "really" have to happen for the dialogue to be meaningful, for the truth of the dialogue resides in what it informs us about our relationship to the world, not in its factuality or accuracy as spoken word. As advanced here, this relationship is participatory and therefore erotic. We live within and are a part of a field of being. Socrates understood this. It was the only thing of which he was certain, he claimed. It follows that as part of the whole, humans are not privy to the meaning of whole. Baudrillard understands this much as well. "It is the nature of meaning that not everything has it," he observes.[5] Precisely. Not everything is amenable to sense making. Moreover, even when meaning is forthcoming, it is always provisional. There are, in short, limits to meaning, just as there are limits to understanding and to our technological powers of control. This is the lesson we moderns have not learned and, as long as we remain modern, are not likely to learn.

The way forward is to find a more salutary way of coping with the existential unease that is part and parcel of the human condition. It is a way that avoids the pitfalls associated with unthinkingness, on the one hand, and the kind of neurosis that develops in response to looking for something that cannot be found, on the other. This alternative is a way that rules out in advance the possibility of attaining anything like certain meaning or understanding. If we could come to terms with the limits of our powers of understanding, as Baudrillard and Camus and others advise we do, the angst that fuels the drive to perfect reality by technological means would lose much of its force. The curse of reality would be replaced by the "great game" of the real, where meaning, always conditional, arises in our participating in an order that transcends the boundaries of the recreated real. We would come to appreciate that the only meaning we are privy to is the meaning that accrues through playing with and being played by a seductive world. We would understand that meaning, such as it is, lies in the searching, not the found, in the doing, not the "how to" do, in performing those deeds apps cannot do for you.

I have operated throughout on the assumption that the eschatological dimension of technology needs airing. There is a need to make explicit what lies half-concealed beneath the surface, namely, that technology holds a supernumerary power. It might lead to our salvation or ruination, but either way it is no trifling force in our lives. This much appears true. But if it is true it is only because we have failed to question seriously the endpoint implied in the ceaseless drive to mastery and its bearing on us. The conceit of this study is that thinking about technology exposes our devotion to the ideal of technological mastery in a way that undermines it.

My hope is that finding the language and the corresponding means to think about technology will lead to taking whatever measures we can to forestall what typically counts as progress today. Sometimes the best path is a crooked path. Not doing what technology wants you to do might make your life less efficient, but what you will gain from your diversions will far outweigh any inconveniences.

NOTES

1. The so-called "end of history" debate surfaced in modern times with the publication of Alexandre Kojève's *Introduction to the Reading of Hegel: Lectures on the "Phenomenology of Spirit,"* ed. Allan Bloom, tr. James H. Nichols (Ithaca, NY: Cornell University Press, 1980). More recent texts dealing with our post-historical condition include Tom Darby's *The Feast: Meditations on Politics and Time* (Toronto: The University of Toronto Press, 1982), and Francis Fukuyama's *The End of History and the Last Man* (New York: Free Press, 2006).

2. In a 1966 *Der Spiegel* interview Heidegger was asked whether "the individual man" or "philosophy" can alter or influence our technological fate. His response was this: "If I may answer briefly, and perhaps clumsily, but after long reflection: philosophy will be unable to effect any immediate change in the current state of the world. This is true not only of philosophy but of all purely human reflection and endeavor. Only a god can save us. The only possibility available to us is that by thinking and poetizing we prepare a readiness for the appearance of a god, or for the absence of a god in [our] decline, insofar as in view of the absent god we are in a state of decline." The quotation above is retrieved from the online site: http://www.ditext.com/heidegger/interview.html

3. The phrase "acting into nature" is drawn from Barry Cooper's *Action into Nature: An Essay on the Meaning of Technology* (Notre Dame, IN: University of Notre Dame Press, 1991).

4. Plato, *Phaedrus,* 230b.

5. Jean Baudrillard, *The Intelligence of Evil: Or the Lucidity Pact,* tr. Chris Turner (London: Bloomsbury, 2013), 13. The full passage from which this extract is taken, reads, "What I call Integral Reality is the perpetrating on the world of an unlimited operational project whereby everything becomes real, everything becomes visible and transparent, everything is 'liberated,' everything comes to fruition and has meaning (whereas it is in the nature of meaning that not everything has it)."

Bibliography

Angell, Ian O. and Demetis, Dionysios S. *Science's First Mistake: Delusions in Pursuit of Theory.* London: Bloomsbury Academic, 2010.

Arendt, Hannah. *Between Past and Future: Eight Exercises in Political Thought.* New York: Viking, 1961.

Aristotle. *Nicomachean Ethics.* Translated by Martin Ostwald. Indianapolis: The Bobbs-Merrill Company, 1962.

Bacon, Francis. *The Oxford Francis Bacon.* Oxford: Clarendon Press, 2008.

Badmington, Neil. *Alien Chic: Posthumanism and the Other Within.* London: Routledge, 2004.

Ball, Terence, Dagger, Richard, and O'Neill, Daniel. *Ideals and Ideologies: A Reader.* 9th ed. Toronto: Pearson Education, 2014.

Barney, Darin. *The Network Society.* Cambridge, UK: Polity, 2004.

Baudrillard, Jean. *The Agony of Power.* Translated by Ames Hodges. Chris Turner. Los Angeles: Semiotext(e), 2010.

———. *Carnival and Cannibal.* Translated by Chris Turner. London: Seagull Books, 2010.

———. *The Conspiracy of Art.* Translated by Ames Hodges. Edited by Sylvère Lotringer. New York: Semiotext(e), 2005.

———. *The Illusion of the End.* Translated by Chris Turner. Stanford, CA: Stanford University Press, 1994.

———. *The Intelligence of Evil or the Lucidity Pact.* Translated by Chris Turner. London: Bloomsbury, 2013.

———. *Passwords.* Translated by Chris Turner. London: Verso, 2003.

———. *The Perfect Crime.* Translated by Chris Turner. London: Verson, 2008.

———. *Simulacra and Simulation.* Translated by Sheila Faria Glaser. (Ann Arbor: University of Michigan Press, 1995.

———. *The Transparency of Evil: Essays on Extreme Phenomena.* Translated by James Benedict, London: Verso, 1993.

———. *The Vital Illusion.* Edited by Julia Witwer. New York: Columbia University Press, 2000.

———. *Why Hasn't Everything Already Disappeared?* Translated by Chris Turner. London: Seagull Books, 2009.

Benedikt, Michael, ed. *Cyberspace: First Steps.* Cambridge, MA: The MIT Press, 1991.

Bohm, David. *Wholeness and the Implicate Order.* London: Routledge Classics, 2002.

Borgmann, Albert. *Crossing the Postmodern Divide.* Chicago: The University of Chicago Press, 1992.

———. *Holding On to Reality: The Nature of Information at the Turn of the Millennium.* Chicago: The University of Chicago Press, 1999.

————. *Technology and the Character of Contemporary Life: A Philosophical Inquiry*. Chicago: The University of Chicago Press, 1984.

Braidotti, Rosi. *The Posthuman*. Cambridge, UK: Polity, 2013.

Brown, Steven, T. *Tokyo Cyberpunk: Posthumanism in Japanese Visual Culture*. Basingstoke, UK: Palgrave Macmillan, 2010.

Buber, Martin. *I and Thou*. Translated by Ronald Gregor Smith. Edinburgh: T. & T. Clarke, 1947.

Camus, Albert. *Lyrical and Critical Essays*. Translated by Ellen Conroy Kennedy. Edited by Philip Thody. New York: Vintage Books, 1968.

Carr, Nicholas. *The Glass Cage: How Our Computers Are Changing Us*. New York: W. W. Norton, 2015.

————. *The Shallows: What the Internet Is Doing to Our Brains*. New York: W. W. Norton, 2011.

————. *Utopia Is Creepy: And Other Provocations*. New York: W. W. Norton, 2016.

Chesterton, G.K. *Orthodoxy*. New York: Dodd, Mead & Co., 1908.

Cooper, Barry. *Action Into Nature: An Essay on the Meaning of Technology*. Notre Dame, IN: University of Notre Dame Press, 1991.

Crary, Jonathan. *24/7: Late Capitalism and the Ends of Sleep*. London: Verso, 2013.

Crawford, Matthew B. *The World Beyond Your Head: On Becoming an Individual in an Age of Distraction*. New York: Allen Lane, 2015.

————. *Shop Class As Soulcraft: An Inquiry into the Value of Work*. New York: Penguin Books, 2010.

Darby, Tom. *The Feast: Meditations on Politics and Time*. Toronto: University of Toronto Press, 1982.

Descartes, René. *The Philosophical Writings of Descartes: Volume 1*. Translated by John Cottingham, Dugald Murdoch, and Robert Stoothoff. Cambridge: Cambridge University Press, 1985.

Diamandis, Peter and Kotler, Steven. *Abundance: The Future Is Better Than You Think*. New York: Free Press, 2014.

Dodds, E. R. *The Greeks and the Irrational*. Berkeley: University of California Press, 1951.

Dreyfus, Hubert. *On the Internet*. Second edition. London: Routledge, 2009.

Dreyfus, Hubert and Kelly, Sean Dorrance. *All Things Shining: Reading the Western Classics to Find Meaning in a Secular Age*. New York: Free Press, 2011.

Eco, Umberto. *Foucault's Pendulum*. Translated by William Weaver. New York: Harcourt Brace Jovanovich, 1989.

Einstein, Albert and Infield, Leopold. *The Evolution of Physics: From Early Concepts to Relativity and Quanta*. New York: Simon and Schuster, 1938.

Ferry, Luc. *A Brief History of Thought: A Philosophical Guide to Living*. Translated by Theo Cuffe. New York: Harper Perennial, 2011.

Feyerabend, Paul. *Conquest of Abundance: A Tale of Abstraction Versus the Richness of Being*. Edited by Bert Terpstra. Chicago: The University of Chicago Press, 2001.

————. *The Tyranny of Science*. Edited by Eric Oberheim. Cambridge, UK: Polity, 1996.

Feynman, Richard. *Surely You're Joking, Mr. Feynman! (Adventures of a Curious Character)*. Edited by Edward Hutchings. New York: W. W. Norton, 1997.

Franklin, Ursula. *The Real World of Technology*. Montreal: CBC Enterprises, 1990.

Frisch, Max. *Homo Faber*. Translated by Michael Bullock. Fort Washington, PA: Harvest Books, 1994.

Fukuyama, Francis. *Our Posthuman Future: Consequences of the Biotechnology Revolution*. New York: Picador, 2003.

Genesko, Gary, ed. *The Uncollected Baudrillard*. London: Sage, 2001.

Germain, Gil. *A Discourse on Disenchantment: Reflections on Politics and Technology*. Albany: The State University of New York Press, 1993.

————. *Spirits in the Material World: The Challenge of Technology*. Lanham, MD: Lexington Books, 2009.

Grant, George. *George Grant in Process: Essays and Conversations*. Edited by Larry Schmidt. Toronto: Anansi, 1978.

———. *Technology and Empire: Perspectives on North America.* Toronto: Anansi, 1969.

———. *Technology and Justice.* Toronto: Anansi, 1986.

Gray, John. *Black Mass: Apocalyptic Religion and the Death of Utopia.* New York: Farrar, Straus, and Giroux, 2008.

Habermas, Jürgen. *The Theory of Communicative Action,* Vol. 1. Translated by Thomas McCarthy. Boston: Beacon Press, 1984.

———. *Toward a Rational Society: Student Protest, Science, and Politics.* Translated by Jeremy J. Shapiro. Boston: Beacon Press, 1970.

Hale, Kimberly Hurd. *Francis Bacon's New Atlantis in the Foundation of Modern Political Thought.* Lanham: Lexington Books, 2013.

Harris, Michael. *The End of Absence: Reclaiming What We've Lost in a World of Constant Connection.* Toronto: HarperCollins, 2014.

Hayles, N. Katherine. *How We Became Posthuman: Virtual Bodies in Cybernetics, Literature, and Informatics.* Chicago: The University of Chicago Press, 1999.

Heath, Joseph. *The Efficient Society: Why Canada Is As Close To Utopia As It Gets.* Toronto: Penguin Canada, 2001.

Heidegger, Martin. *The Question Concerning Technology and Other Essays.* Translated by William Lovitt. New York: Harper Colophon Books, 1977.

———. *What Is Called Thinking?* Translated by Glenn Gray. New York: Harper Colophon Books, 1968.

Heim, Michael. *The Metaphysics of Virtual Reality.* Oxford: Oxford University Press, 1993.

Herbrechter, Stefan. *Posthumanism: A Critical Analysis.* London: Bloomsbury Academic, 2013.

Hobbes, Thomas. *The Leviathan.* Buffalo: Prometheus Books, 1988.

Houellebecq, Michel. *The Possibility of an Island.* New York: Vintage, 2007.

Ihde, Don. *Bodies in Technology.* Minneapolis: University of Minnesota Press, 2001.

———. *Technics and Praxis.* Dordrecht: D. Reidel, 1979.

———. *Technology and the Lifeworld: From Garden to Earth.* Indianapolis: University of Indiana Press, 1990.

Ishiguro, Kazuo. *Never Let Me Go.* New York: Vintage, 2006.

Jackson, Maggie. *Distracted: The Erosion of Attention and the Coming Dark Age.* Amherst: Prometheus Books, 2008.

Kellmereit, Daniel and Obodovski, Daniel. *The Silent Intelligence: The Internet of Things.* Warwick, NY: DND Ventures, 2013.

Kelly, Kevin. *What Technology Wants.* New York: Penguin Books, 2011.

———. *The Inevitable: Understanding the 12 Technological Forces That Will Shape Our Future.* New York: Viking, 2016.

Kohák, Erazim. *The Embers and the Stars: A Philosophical Inquiry into the Moral Sense of Nature.* Chicago: The University of Chicago Press, 1984.

Kojève, Alexandre. *Introduction to the Reading of Hegel: Lectures on the Phenomenology of Spirit.* Translated by James H. Nichols, Jr. Edited by Allan Bloom. Ithaca, NY: Cornell University Press, 1969.

Koosed, Jennifer, ed. *The Bible and Posthumanism.* Atlanta: Society of Biblical Literature, 2014.

Kotler, Steven. *Tomorrowland: Our Journey from Science Fiction to Science Fact.* Seattle: New Harvest, 2015.

———. *The Rise of Superman: Decoding the Science of Ultimate Human Performance.* Seattle: New Harvest, 2014.

Kundera, Milan. *The Unbearable Lightness of Being.* Translated by Michael Henry Heim. New York: Harper Perennial Modern Classics, 2005.

Kurzweil, Ray. *How to Create a Mind: The Secret of Human Thought Revealed.* New York: Viking, 2012.

———. *The Singularity Is Near: When Humans Transcend Biology.* New York: Viking, 2006.

Locke, John. *Second Treatise of Government.* Edited by C.B. Macpherson. Indianapolis: Hackett, 1980.

Lyotard, Jean-Francois. *The Inhuman: Reflections on Time.* Stanford, CA: Stanford University Press, 1991.

Machiavelli, Niccolò. *The Prince.* Second edition. Translated by Harvey C. Mansfield. Chicago: The University of Chicago Press, 1998.

Marx, Karl, with Engels, Friedrich. *The Communist Manifesto.* Edited by Frederic L. Bender. New York: W. W. Norton, 1988.

———. *The German Ideology, Including Theses on Feuerbach and the Introduction to the Critique of Political Economy.* Amherst, MA: Prometheus Books, 1998.

McAfee, Andrew. *The Second Machine Age: Work, Progress, and Prosperity in a Time of Brilliant Technologies.* New York: W. W. Norton, 2014.

McCarthy, Cormac. *The Road.* New York: Vintage, 2006.

McKnight, Stephen A. *Sacralizing the Secular: The Renaissance Origins of Modernity.* Baton Rouge: Louisiana State University Press, 1989.

Merleau-Ponty, Maurice. *Phenomenology of Perception.* Translated by Colin Smith. London: Routledge and Kegan Paul, 1962.

———. *The Primacy of Perception: And Other Essays on Phenomenological Psychology, the Philosophy of Art, History, and Politics.* Edited by James M. Edie. Evanston, IL: Northwestern University Press, 1964.

———. *The Visible and the Invisible.* Translated by Alphonso Lingis. Edited by Claude Lefort. Evanston, IL: Northwestern University Press, 1968.

Monod, Jacques. *Chance and Necessity: An Essay of the Natural Philosophy of Modern Biology.* Translated by Austryn Wainhouse. New York: Alfred A. Knopf, 1971.

Musil, Robert. *The Man Without Qualities Vol. 1: A Sort of Introduction and Pseudo Reality Prevails.* Translated by Sophie Wilkins. Edited by Burton Pike. New York: Vintage, 1996.

Nagel, Thomas. *Mind and Cosmos: Why the Materialist Neo-Darwinian Conception of Nature Is Almost Certainly False.* Oxford: Oxford University Press, 2012.

Nayar, Pramod, K. *Posthumanism.* Cambridge, UK: Polity, 2013.

Nietzsche, Friedrich. *The Basic Writings of Friedrich Nietzsche.* Translated by Walter Kaufmann. New York: The Modern Library, 1992.

———. *Thus Spoke Zarathustra.* Translated by R. J. Hollingdale. New York: Penguin, 1961.

———. *The Will to Power.* Translated by Walter Kaufmann and R.J. Hollingdale. New York: Vintage, 1968.

Noble, David F. *The Religion of Technology: The Divinity of Man and the Spirit of Invention.* New York: Alfred A. Knopf, 1998.

Noë, Alvin. *Out of Our Heads: Why You Are Not Your Brain, and Other Lessons from the Biology of Consciousness.* New York: Hill and Wang, 2009.

O'Connor, R. Eric. *Conversations With Eric Voegelin.* Montreal: Thomas More Institute, 1980.

Orwell, George. *Nineteen Eighty-Four.* London: Penguin Books, 1989.

Pentland, Alex. *Social Physics: How Social Networks Can Make Us Smarter.* New York: Penguin Books, 2015.

Phillips, Adam. *Missing Out: In Praise of the Life Unlived.* New York: Farrar, Straus and Giroux, 2012.

Plato. *The Collected Dialogues of Plato.* Edited by Edith Hamilton and Huntington Cairns. Princeton, NJ: Princeton University Press, 1961.

Poerksen, Uwe. *Plastic Words: The Tyranny of a Modular Language.* Translated by Junta Mason and David Cayley. University Park: The Pennsylvania State University Press, 1995.

Robinson, Marilynne. *Absence of Mind: The Dispelling of Inwardness from the Modern Myth of the Self.* New Haven, CT: Yale University Press, 2011.

———. *The Givenness of Things.* New York: HarperCollins, 2015.

———. *When I Was A Child I Read Books.* New York: Harper Perennial, 2013.

Rossi, Paolo. *Francis Bacon: From Magic to Science.* Chicago: The University of Chicago Press, 1978.

Saint-Simon, Henri. *The Political Thought of Saint-Simon.* Translated by V. Ionescu. Edited by Ghita Ionescu. Oxford: Oxford University Press, 1976.

Saunders, George. *In Persuasion Nation.* New York: Riverhead Books, 2007.

Segalen, Victor. *Essay on Exoticism: An Aesthetics of Diversity.* Translated and edited by Yaël Rachel Schlick. Durham, NC: Duke University Press, 2002.

Shirky, Clay. *Cognitive Surplus: How Technology Makes Consumers into Collaborators.* New York: Penguin Books, 2011.

Slouka, Mark. *War of the Worlds:Cyberspace and the High-Tech Assault on Reality.* New York: BasicBooks, 1995.

Srigley, Ronald D. *Albert Camus' Critique of Modernity.* Columbia: University of Missouri Press, 2011.

Tabachnick, David. *The Great Reversal: How We Let Technology Take Control of the Planet.* Toronto: University of Toronto Press, 2013.

Tabachnick, David and Koivukoski, Toivu, eds. *Globalization, Technology, and Philosophy.* Albany: State University of New York Press, 2004.

Taras, David. *Digital Mosaic: Media, Power, and Identity in Canada.* Toronto: University of Toronto Press, 2015.

Taylor, Charles. *The Malaise of Modernity.* Concord: Anansi, 1991.

Tenner, Edward. *Why Things Bite Back: Technology and the Revenge of Unintended Consequences.* New York: Alfred A. Knopf, 1996.

Tuan, Yi-Fu. *Space and Place: The Perspective of Experience.* Minneapolis: University of Minnesota Press, 1977.

Turkle, Sherry. *Alone Together: Why We Expect More from Technology and Less from Each Other.* New York: Basic Books, 2012.

———. *Life On the Scree: Identity in the Age of the Internet.* New York: Simon and Schuster, 1995.

———. *Reclaiming Conversation: The Power of Talk in a Digital Age.* New York: Penguin, 2015.

Voegelin, Eric. *Anamnesis.* Translated by Gerhart Niemeyer. (Notre Dame, IN: University of Notre Dame Press, 1978.

———. *The New Science of Politics.* Chicago: The University of Chicago Press, 1952.

Weber, Max. *From Max Weber: Essays in Sociology.* Translated by Hans Gerth and C. Wright Mills. New York: Oxford University Press, 1946.

———. *The Protestant Ethic and the Spirit of Capitalism.* Translated by Talcott Parsons. New York: Charles Scribner's Sons, 1958.

White, Kenneth. *The Wanderer and his Charts.* Edinburgh: Polygon, 2004.

Wolfe, Cary. *What Is Posthumanism?* Minneapolis: University of Minnesota Press, 2009.

Wooley, Benjamin. *Virtual Worlds: A Journey in Hype and Hyperreality.* Oxford: Blackwell, 1992.

Zamyatin, Yevgeny. *We.* Translated by Clarence Brown. New York: Penguin Twentieth Century Classics, 1993.

de Zengotita, Thomas. *Mediated: How the Media Shapes Your World and the Way You Live in It.* New York: Bloomsbury, 2005.

Žižek, Slavoj. *Event: Philosophy in Transit.* London: Penguin, 2014.

———. *Violence: Six Sideways Reflections.* New York: Picador, 2008.

Index

About the Author

Gil Germain is a professor of political thought in the University of Prince Edward Island's department of political science. His previously published books include *A Discourse on Disenchantment: Reflections on Politics and Technology* (SUNY, 1993) and *Spirits in the Material World: The Challenge of Technology* (Lexington Books, 2009).